Indra's Net and the Midas Touch

Indra's Net and the Midas Touch

Living Sustainably in a Connected World

Leslie Paul Thiele

The MIT Press
Cambridge, Massachusetts
London, England

For information about special quantity discounts, please e-mail special_sales@ mitpress.mit.edu

This book was set in Stone Serif and Stone Sans by Toppan Best-set Premedia Limited. Printed and bound in the United States of America.

Library of Congress Cataloging-in-Publication Data

Thiele, Leslie Paul.
Indra's net and the Midas touch : living sustainably in a connected world / Leslie Paul Thiele.
    p.   cm.
Includes bibliographical references and index.
ISBN 978-0-262-01609-4 (hardcover: alk. paper)
1. Sustainable development. 2. Globalization—Environmental aspects.
3. Globalization—Social aspects. 4. Environmental degradation—Social aspects.
5. Environmental policy. I. Title.
HC79.E5T4758   2011
338.9′27—dc22

                                                                2010054272

10 9 8 7 6 5 4 3 2 1

Contents

Preface

Climate change may well be the greatest challenge that humankind has ever faced. The stakes are certainly the highest imaginable: civilization itself is threatened, and the menace is imminent. We appear to be hurtling toward a tipping point. Maybe we have already passed it. Beyond this dire marker, measured in parts per million of carbon dioxide in the atmosphere, our best efforts of remediation would fail to reverse, or even much temper, a self-reinforcing and accelerated cooking of our planet. Whether we are at a cataclysmic point of no return, or simply worsening an already massive problem, one thing is clear: greater and swifter action is required.

For all the attention that climate change has gained as of late, we still fail to understand its fundamental nature. The problem is not that our best climatologists have miscalculated. They have done their jobs well enough. The warming of the planet, scientists have demonstrated beyond a reasonable doubt, is largely the product of human enterprise, primarily the burning of fossil fuels and the destruction of forests. The widespread acknowledgment that global warming is anthropogenic in nature, although tardy, is a welcome development. But in acknowledging our role as the engineers of the greatest change to the earth's atmosphere that has occurred in the past half-million years, we obscure at least as much as we reveal. By dutifully accepting the burden of responsibility, we have also reinforced the traditional image of humankind as the most potent force on the planet. Yet this image grossly misrepresents the nature of our power and makes it ever more likely that more menacing problems will appear on the horizon.

We have effected more change on earth than any other single species. In this respect, *Homo sapiens* is indeed a powerful planetary force. But the potentially irreversible alteration of natural cycles and planetary conditions has been the unintended consequence of our actions. We did not set out to overrun the globe by our sheer numbers, depriving other species, and ourselves, of adequate resources and habitat. Human population growth beyond the carrying capacity of our planet is the by-product of economic and technological developments. We did not set out to poison our lands, lakes, skies, and seas. Pollution is a side effect of our industrial way of life. We did not set out to deplete the stratospheric ozone layer that protects organic life from ultraviolet radiation. It is an unintended consequence of our invention and commercial production of chemicals. And we did not set out to warm our planet under a blanket of greenhouse gases. It is the unforeseen result of our exploitation of fossil fuels and forests. In short, the most menacing and pressing problems that we face today are by-products. We do not fully appreciate the fact that climate change is simply the latest, albeit most dire, of a long list of unintended consequences.

Climate change confronts us as an unparalleled crisis. Optimists are wont to suggest that it also presents us with an unprecedented opportunity, and they are right. The problem of climate change may force us as a species to desist from our constant quibbling and tribal jousting so that we might mount a truly cooperative global response. The magnitude of this challenge, and opportunity, cannot be overstated. Yet in an important sense, it pales in comparison to the challenge of acknowledging and abiding by what has been called the first law of human ecology. This law, powerful in its simplicity, states that we can never do merely one thing. Every action has unintended consequences, and the larger the scope of the action, the more momentous are its side effects. If we respond to the crisis of climate change in the same old fashion that the species has confronted other problems—as a predicament to be bested by human ingenuity—we will be committing a deadly error. Such an approach will simply reinforce the patterns of thought and action that brought us to this precipice. Our current crisis is unprecedented in its potential for catastrophe, and it is rife with unknowns. But there is one thing we do know with certainty: climate change is not the first and will not be the last instance of a cataclysm produced by the unintended consequences of our craft.

Fundamentally transforming our patterns of thinking and habits of behavior in light of this fact is the real challenge today. Ignoring the need of such transformation is the real danger. We can and must be ingenious and innovative in our practical efforts to address climate change. Meeting this challenge in the absence of a more fundamental transformation, however, will be remembered not as humanity's finest hour but as the most costly lesson left unlearned.

Doom and gloom are easy to declare and hard to dispel. In the face of daunting challenges to our common sense, our practical welfare, and our spiritual well-being, this book fosters an awareness and appreciation of the connectedness of life. It is meant to serve as a resource for those who want to avoid despondency in the face of upheaval. It is an expression of hope—for creativity and community—in a time of danger.

Acknowledgments

I am grateful to my undergraduate students in Facets of Sustainability and Sustainability in Action whose dedicated citizenship—local, national, and global—is changing the world. Their insights and efforts are inspiring. Special thanks are due to my friend and colleague Paul Wapner, to John Klauder, and to my graduate students Evgenia Ilieva, Chris Manick, and Seaton Tarrant for their suggestions and critical commentary. I am also very grateful to B. C. Nelson, from whom I have learned much about community. To my wife, Susan, and sons, Jacob and Jonah, my gratitude, blessedly, has opportunity to be voiced each day.

Introduction: The Fabric of Life

In the beginning is the relation.
—Martin Buber, *I and Thou*

Consider the unnerving features of contemporary life: global climate change, massive species extinctions, acute shortages of freshwater and other natural resources, growing food insecurity, impoverishment and economic devastation, increasing disparities between rich and poor, political instability and alienation, rising numbers of failed states, the juggernaut of genetically modified environments and unprecedented technological transformations. Nearly 7 billion human beings are trying to make ends meet on a finite planet, and these numbers will grow substantially before they begin to decline. An increasing portion of this population is committed to the individual and collective pursuit of endless and accelerating economic growth and technological mastery. The effects of all this enterprise are world changing. Threats to prosperity—indeed, to civilization itself—surround us.

These threats are not distant prospects. They have a direct impact on us today, and we have an impact on them. Each day our choices bear on the sustainability of our lifestyles, communities, and planet. There are no shortages of dilemmas: paper or plastic bags at the grocery checkout; cloth or disposable diapers for our babies; e-readers or paperback books; blue-collar jobs or wilderness preservation; open immigration or tighter borders; nuclear power in the face of rapid climate change or the slow development of renewable forms of energy, such as wind farms that may kill large numbers of bats and birds. The decisions we make as consumers and citizens will have significant consequences. Yet clear answers to the multiple dilemmas we face each day remain elusive and much in dispute by experts.

In the pages that follow, I argue that there are two fundamental reasons for the growing list of dilemmas we encounter in the pursuit of sustainability. First, our globalizing world is increasingly characterized by webs of interdependence. Never before have so many people, activities, and events been so closely connected. Second, and as a consequence of these expanding and deepening interdependencies, the law of unintended consequences has asserted its jurisdiction not only in the domain of ecology but across various fields of inquiry and facets of life.

The growth of social, political, technological, and economic interdependence around the planet, commonly known as globalization, has been well documented over the past four decades.[1] It is not in doubt. Since the era of the New Deal, social scientists have also charted the unintended consequences of economic and social policy. With the rise of environmental concerns in the 1970s, and increasingly in recent years, natural scientists have documented the unintended effects of human enterprise on ecosystems. Still, an interdisciplinary account of expanding and deepening global interdependencies across diverse fields of inquiry and facets of life is lacking, as is an account of the chief effect of these complex connections: the growing jurisdiction of the law of unintended consequences. *Indra's Net and the Midas Touch* provides such an account.

As webs of interdependence increase in size and complexity, so does our incapacity to control the effects of our actions. Some fields of inquiry and endeavor can, and occasionally do, find reassurance in this intimidating reality. Others are prone to ignore or deny it. What is needed in this context is the cultivation of a certain sensibility, set of values, knowledge, and know-how within and across diverse disciplines. The human ability to understand and navigate the web of life is, at one and the same time, a practical skill, an intellectual capacity, a moral disposition, and a form of mindfulness. I call this much-needed capacity *ecosophic awareness*.[2]

In many respects, ecosophic awareness resembles what the ancient Greeks called *phronesis*: practical wisdom that is also known as prudence. It was understood to constitute both a moral and an intellectual virtue. Practical wisdom is a way of understanding that produces a better way of acting and a way of acting that stimulates deeper understanding. It resembles the concept of *prajna* or wisdom within the Buddhist tradition. *Prajna* allows one to understand the cyclical processes of *karma* (that is, action) and act appropriately in the face of interdependent actions and effects. It also allows a deep and transformative understanding of the nondualistic nature of reality. This nondualism ultimately grounds the processes of *karma*. For Buddhists, wisdom allows an appreciation of the interconnection and inseparability of actions and effects, as well as an appreciation of the interconnection and inseparability of the beings that produce and are produced by these actions and effects. Although I draw on both Western and Eastern traditions, this book is neither designed nor intended as an account of ancient Greek philosophy, Buddhist thought, or any other canon. Rather, it employs the term *ecosophic awareness* to designate a sensibility fit for the daunting challenges and deep complexities of this century.

Ecosophic awareness might best be defined as a sage appreciation of the ubiquity of interdependence combined with the disposition toward contextually responsive engagement. It provides the intellectual and moral foundation for efforts to sustain the web of life in a world of unintended consequences. My use of the term is meant to call to mind its etymology.

The word *ecology* (*oekologie*) was coined in 1873 by the German zoologist Ernst Haeckel. It referred then, as it does now, to the study (*logos* means word, reason, or discourse) of the interactive relations of plants and animals in their natural habitats. The *eco-* in *ecology* comes from the Greek term *oikos*, which designated a household or dwelling place. The dwelling places of concern to ecologists are the various habitats that are used, maintained, and transformed by the organisms that occupy them. Ecologists are concerned with relationships of interdependence. In fact, the first uses of the term *interdependence* in reference to natural phenomena occurred in the 1870s and 1880s just as ecology was developing as a discipline.[3] To the extent that ecologists focus on particular organisms, they do so insofar as these organisms live within, contribute to, and depend on vast and intricate biophysical networks. Relationships are primary and fundamental. As Theodore Roszak succinctly states, "Ecology is the study of connectedness."[4]

To be *oikos*-sophic, then, is to be prudently engaged with the nested habitats of complex interdependence that constitute the web of life. In moral and political life, no less than in ecological systems, nothing issues from a single cause, generates a single effect, or has a single meaning. To be *oikos*-sophic is to understand that there are no solitary causes or effects.

Navigating the world skillfully means responding to context, engaging with contingency, and anticipating unintended consequences. Ecosophic awareness signifies attentiveness and responsiveness to a world in flux. It fosters adaptation to a constantly changing environment.

The term *awareness* recalls the seventh element of the Buddha's eight-fold path to enlightenment. Variously designated as right mindfulness, right memory, right attention, or right awareness, the penultimate element of the noble eightfold path is the practice of staying open to, bringing to attention, and remembering the phenomena that affect mind and body. It also entails bringing to attention the thoughts, beliefs, and motivations that cause us to act on, or react to, our world. The practice of right aware-ness is said to produce wisdom. For my purposes, it designates an attentive-ness to relationships that fosters prudent interactions. These relationships are both external to what we normally designate as the self and constitu-tive of it.

Ecosophic awareness is a mindful attention to relations of interdepen-dence and a hopeful investment in them. Hopefulness is not faith or blind optimism. Rather, it is the capacity to perceive one's world as pregnant with possibilities. To be hopefully engaged in relationships of interdepen-dence means to embrace our roles as midwives of the future. That is a very expansive claim, and one that I affirm as the mandate for sustainability.

The alternative to defining sustainability so expansively is to understand it primarily as a technological solution to a technological problem. This is perhaps its most common interpretation. Here sustainability entails gaining increased efficiency in the use of natural resources and greater foresight in the execution of economic enterprise. If burning fossil fuels that release greenhouse gases is the problem, then green, renewable energy is the sus-tainable solution. If absence of market regulations is the problem, institut-ing better economic incentive and disincentive structures is the solution. Understanding sustainability in this manner, primarily as a feat of physical and socioeconomic engineering, is a mistake. To portray sustainability merely as a technological effort—even one that deftly balances environ-mental preservation with economic development and social welfare—is as shortsighted as it is misguided.

To be sure, we need to engineer green solutions to many threatening problems. Greater efficiencies and better planning are all for the good. But the promise of sustainability is found elsewhere. It pushes us beyond purely technological frames of mind. Sustainability encourages a profound engagement with the human condition. It prompts us—for very practical reasons—to fully explore humanity's role in the web of life. My argument is that the discourses and practices of sustainability encourage us to understand, appreciate, and engage our ethical, technological, economic, political, and psychological lives, as well as the ecological and (meta)physical habitats within which we fashion these lives—as nested realms of complex interdependence.

The ancient wisdom traditions first gave voice to the notion of an interdependent world and the practical wisdom required to navigate it. But such inspiring explorations of what I call ecosophic awareness have been periodic and dispersed. Sadly, human history is defined as much by stubborn blindness to the reality of interdependence as adept recognition. And today, for the first time in the history of the species, the unintended consequences of actions stemming from such blindness threaten civilization itself.

Mindful, hopeful engagement with the nested realms of complex interdependence that define the human condition in this century is our most pressing need. The promise of sustainability, beyond any technological solutions it produces for the very real problems we face, is the fostering of ecosophic awareness. Such awareness allows us to become midwives of the future: to safeguard, guide, and witness the interdependent phenomena of being and becoming. My belief is that we will neither achieve more sustainable societies nor understand the nature of our current challenges if we do not explore and embrace the breadth and depth of our interdependencies. Consequently I take readers to places that other books on sustainability seldom, if ever, venture.

Sustainability in a Changing World

Sustainability may be defined as the quest for ever-greater resilience in an interdependent world. The resilience of an ecological system is its capacity to maintain the number and overall pattern of its relationships (regularized interactions between species) in the face of exogenous shocks and internal shifts. More broadly, resilience is the capacity of a social, cultural, or biological system to adapt to and recover from disturbances and change that threaten to undermine its crucial relationships and values. A resilient system persists in time and space. But it is not static. Indeed, resilience is precisely the capacity of a system to adapt to a world in flux without falling apart.

The ancient (Western) Roman Empire came undone in the fifth century as Germanic tribes and troops took control of the Italian peninsula while far-off lands that Rome formerly ruled claimed their independence. The ancient Aztec Empire ended with the Spanish conquest in the sixteenth century. The Roman and Aztec empires were not sufficiently resilient to withstand military, political, social, and environmental shocks that eventually caused them to collapse. That is not to say that all Romans or Aztecs were killed with the collapse of their empires, that Latin and Nahuatl immediately ceased to be spoken languages, or that certain features of Roman and Aztec culture and government did not survive. However, political, military, and cultural relationships and values that were deemed crucial to these empires failed to persist.

To be resilient, a society or culture must sufficiently adapt to changing circumstances so as not to collapse. At the same time, it must maintain its core relationships and values, lest it cease to be identifiable as *that* particular social or cultural system. In cases where radical change is required to ensure survival, the transformation may be so extensive as to constitute not the adaptation of an existing system but the emergence of a new one. Much depends on the rate of change. Given a long enough span of time, significant change can be accommodated while retaining the threads of continuity. Indeed, one might argue that an overlap in time among a sufficient number of relationships or values, rather than the persistence of any particular (set of) core relationships or values, is what constitutes a resilient system.

The ancient Greek sage Heraclitus (whom we will revisit in later chapters) famously announced that you could never step in the same river twice. A river is in constant flux. Its total volume of water fluctuates owing to weather and season, its banks are constantly eroding and being rebuilt, its underwater contours and currents are protean, its organic inhabitants are forever in movement. Yet the river endures. Of course, rivers are not eternal. They are born and die in geological, if not human, time frames. Still, a river may achieve resilience—but only by way of continuity in the face of incessant change. Societies are like rivers: they can be sustained only if rates of change do not undermine fundamental relationships.

One is reminded of the paradox of Theseus's ship. According to Plutarch, when Theseus returned to ancient Athens from the island of Crete after defeating the bull-headed Minotaur, his ship was preserved as a monument to his heroism. As rot set in over the years, one plank after the other was replaced until none of the original planks remained. The Greek philosophers, as was their wont, debated whether the original ship had been preserved or a wholly new one constructed. At any particular point in time, the ship looked very much the same. But over time, there was nothing substantive of the original vessel that remained. Only the form, not the matter, was preserved.

We might have the same debate about our species and planetary habitat. If we preserve the human race and its home on earth over the next millennium by genetically engineering most life forms on the planet and turning ourselves into cyborgs, beings more mechanical and electronic than organic, would that constitute resilience? Some might argue that so much change with so little continuity is not an example of resilience but the creation of a new and different world. Still, the earth's biosphere—sustained in some shape or form for 3.5 billion years—has undergone tremendous change. For most of this period, the earth was a pretty uninteresting place, zoologically speaking. The oldest multicellular animals—certain sponges, coral, and jellyfish—are less than 1 billion years old. Continuous fossil records of clearly identifiable species, such as the horseshoe crab, go back only 400 million years. But even in the past half-billion years, the level of "turnover" has been remarkable. Over 99 percent of all the species that ever existed on the planet are now extinct. If the earth's biosphere is our model for resilience, then resilience and radical change seem quite compatible.

No component of any living system is everlasting. For that matter, no living system is everlasting. Astrophysicists and astronomers inform us that the earth itself will one day perish in fire when the star we call the sun expands into a Red Giant. In about 5 billion years, our life-giving sun will have increased its radius 200-fold, effectively engulfing the first three planets of the solar system. Long before that, in about 1 billion years, our oceans will start to evaporate, marking the beginning of the end of earth as a biological haven. Soon enough, cosmologically speaking, our planetary home will become a lifeless, desiccated satellite orbiting a fiercely growing star. Then it will melt into a large piece of molten rock. To persist in the long term, its denizens will have to colonize other planets or solar systems.

Notwithstanding popular notions that we must "save the earth," the hard truth is that our planet is time bound. We cannot save the earth anymore than we can destroy it. In its 4.5-billion-year history, the planet has undergone more change, and more drastic change, than humankind could ever induce on its own. The future will be no different from the past in this regard. That is a difficult thought. But we need not become mired in philosophical musings about impermanence or lose ourselves within geological or cosmological time frames. Surely sustaining a web of life that allows for diverse, resilient ecosystems and thriving human societies for the foreseeable future is a task worthy of our greatest efforts.

What is usually meant when people speak of "saving the planet" is preserving the anthropocene—the period of the past 10,000 years of earth history that gave rise to human agriculture and, subsequently, to urbanism and all the trappings of culture. When environmentally oriented people say they want to save the earth, what they really want to save is a high quality of civilized life without at the same time destroying the other species and landscapes that currently share the planet. After all, *sustainability* is not simply a descriptive term for the continuity of a biological system. It is a normative term that describes a good to be sought in the here and now and for the foreseeable future. Sustainability refers to a certain sort of resilience—the resilience we can and should achieve in our time for our own sakes, for the welfare of our progeny, and for the benefit of the other species that make up the web of life. Importantly, sustainability pertains not just to ecological survival but to social, economic, and cultural welfare. A sustainable society integrates the four goods that human beings need to pursue in a balanced fashion to achieve resilience: ecological health and diversity, economic security and opportunity, social equity and empowerment, and cultural creativity and learning.

In an interdependent world, the synergistic pursuit of these four goods maximizes the likelihood that civilization can be sustained in the long term. Specifically, the claim is this: societies that well balance the pursuit of economic prosperity with social justice and environmental caretaking while ensuring intellectual and cultural development will prove resilient. This claim requires empirical validation. There is already good evidence for it, and experiments to validate it further are the most reasonable and prudent we can pursue.

The Way Forward

Unless you are a teenager with very good genes, within four-score years you most certainly will be dead. Take a moment to ingest this fact. Nothing that you do in your life, however inspiring, can help you avoid this inevitable conclusion. Soon enough, all of us will be little more than memories. To be sure, many will have left concrete legacies: children and grandchildren, inventions, institutions, inspiring words and actions. These influences and achievements may endure for generations, even centuries. Still, for most of us, one of the most enduring legacies we leave behind will be our contribution to the depletion of the planet's natural resources.

In an average lifetime, individuals like you and me will have consumed—directly by what we eat, drink, buy, and throw away, and indirectly by way of the inefficiencies of industrial production that chew up natural resources to generate the goods we demand—millions of pounds of the planet per person. Owing to the inefficiencies of the system, only 6 percent of these materials end up in the actual products we use, with the rest discarded in processes of production. Over a lifetime, the average American generates more than 130,000 pounds of trash directly, and a great deal more than that indirectly, by way of his or her participation in an industrial economy.[5] Most troubling, however, are not the mountains of trash we create but the vast depletion of natural resources that our consumption and waste represent. Some of these depletions are never to be remedied, as occurs whenever another species goes extinct. As troubling is the depletion of the planet's capacity to absorb the by-products of human productivity. In a lifetime, the average American will have deposited 320,000 pounds of carbon dioxide into the atmosphere, five times the global average.[6] As we have now become painfully aware, the planet's capacity to absorb these greenhouse gases without large-scale alterations of climate has already been exceeded. We will not be around in four score years to feel the effects of our actions. But our great-grandchildren certainly will. They may know us best by neither personal stories passed through the generations nor inventions and achievements that have endured the test of time, but by way of the massive ecological debt and devastation they inherit from us.

Some months ago, I was listening to a podcast on climate change. It featured the CEO of a large corporation who, environmentally speaking, had recently found religion. He was now engaged in an effort to model sustainable business practices at his place of work and carry the good word to fellow executive officers. The transformative moment for the CEO had occurred, innocently enough and without warning, while he was sitting around the dinner table with his family. The mealtime conversation drifted to the topic of global warming, an issue his son had learned about that morning in middle school. Dutifully, the parents offered a nonalarmist but frank assessment of the scope of the problem. After carefully listening to the adults and her older sibling weigh in on the topic, the seven-year-old daughter asked a question that would eventually rock a corporation. "Daddy," the little girl queried, "what are you doing to keep the world from getting too hot?"

The father was stymied. Somehow an extended monologue on the challenges facing businesses in a competitive global marketplace seemed irrelevant. After a long pause and a forkful of steaming peas that resisted swallowing, the CEO sheepishly responded: "Not enough, sweetie. Not enough." That admission was the beginning of a personal, and corporate, transformation.

We might all answer the child's question in like fashion. Today virtually everything we consume or construct taxes the planet's resources and constrains the future of our progeny. To the extent that we directly or indirectly play a role in economies fired by fossil fuels, we cannot avoid aiding and abetting what future generations might deem environmental crimes. These days *mea culpa* goes without saying.

Locked in—as most of us are—to a global marketplace of goods and services, and participating in a global village of communication and interaction, it is difficult, if not impossible, to monitor the multiple ways in which our personal lifestyles, our consumer choices, our business and professional enterprises, and our political actions have an impact on the natural and social world. Figuring out how to make each of these innumerable deeds—or even the lion's share of them—promote rather than undermine sustainability is a daunting challenge. At times, it seems more the prerogative of a deity than a mandate for mortals.

Part of the problem is that thinking through the ramifications of our actions is difficult. The chain of causation is simply too long and twisted. If we attempt to specify the social and environmental effects of action a on b and b on c, we are likely to lose focus before we get to f or g, let alone z. Indeed, the challenge is more profound. Living as we do in networks of interdependence, action a does not simply produce effect b. It also produces side effect b^1 or, more likely, side effects b^1, c^1, and d^1. Each of these side effects in turn serves as cause for another series of effects and then side effects. Like a stone thrown in a pond, the repercussions of our actions ripple out like waves in all directions. Each of these waves, when encountering an obstacle in its path, produces a new set of waves radiating with altered frequencies and amplitudes. The mind reels at the possibilities. Even the impact of the least of our actions is beyond our capacity to compute or comprehend. The imperative to live sustainably within the vast and intricate web of interdependent relationships that constitutes our world would appear to require knowledge verging on omniscience and power just shy of omnipotence.

In this respect, sustainability is like fine sand. It is easy enough to hold in a loosely cupped hand, and it can be used to build enduring structures. But the more one tightens one's grip, the more it slips through one's fingers. As an ideal, sustainability is easy enough to grasp. As a principle and vision, it can be employed to build lasting communities and ecologies. When we attempt to squeeze it too tightly, however, with the intent of crafting comprehensive policies and permanent prescriptions, it escapes our grasp. The mark of education and culture, Aristotle observed, is the pursuit of only as much precision in a subject as its nature permits. Sustainability permits neither precise prognosis nor rigid policy. It is a dance with uncertainty.

As a subject for study, sustainability is inherently interdisciplinary. It requires navigating connections and managing interactions between diverse human enterprises. We live in an increasingly specialized world where a professional must know more and more about less and less simply to keep abreast of the accelerating developments within her field of expertise. Yet our world is in great need of cross-disciplinary inquiry and integrated practice. Sustainability requires synthesis. It demands the creative engagement with an ever-broadening community of stakeholders and the adaptive management of dynamic relationships, interdependent sets of issues, and unintended consequences.

To say that sustainability is a dance with uncertainty is not to sanction the cultivation or exploitation of ignorance. (The systematic effort by sections of the fossil fuel industry and its political allies to undermine public understanding of the science of climate change is perhaps the most egregious example.) There is enough uncertainty and ignorance in the world to go around. The last thing we need are self-serving campaigns of misinformation aimed at sowing doubt and passivity. And to say that sustainability does not truck stubborn formulations and permanent prescriptions is not to sanction empty commitments or lukewarm efforts. Not at all. The point is simply that sustainability is not a theoretical enterprise aimed at closure; it is an iterated practical exercise. Though well grounded in principles, sustainability—like justice, liberty, or any other ideal—does most of its work through the contested exploration of its meaning and the tentative yet concrete embodiments of its pursuit. Like justice and liberty, sustainability is a fruit of Tantalus. It forever escapes our grasp and, for that very reason, extends our reach.

Although essentially contested in its meanings and impossible to attain in any absolute or unchanging form, sustainability presents itself as an imperative. Clearly business as usual is not an option, at least not if we value civilization and the diversity of life. At the current rate of demographic increase, economic growth, and technological expansion, we would need to colonize many more planets to maintain our current trajectory of consuming, disrupting, or despoiling clean air, freshwater, arable land, fossil resources, wilderness, biodiversity, and a stable climate. And if our current practices are any indication, these newly colonized planets, like the one we now call home, would be marked by astounding levels of inequity and injustice. But if we cannot carry out business or our personal lives as usual in the face of such monumental concerns, then what should be done?

Many books available today attempt to answer this crucial query by endorsing a particular public policy or supplying a wish list of technological innovations. *Indra's Net and the Midas Touch* is not one of them. It does not identify specific policies to adopt or specific things to build or buy. This book does not supply the silver bullets that people understandably hope to gather at political rostrums, discover in laboratories, or pluck off store shelves. In the arena of sustainability, there are no silver bullets. Remedies touted as such inevitably ricochet.

The point is not to abandon hope. Quite the contrary. Hope is our greatest resource in these troubling times. But the hope we claim and cultivate must come from decidedly new ways of thinking and acting. My intent is to provide a means for readers to reorient their lives from the vantage point of a new set of nested habitats. From this sobering yet invigorating vantage point, we can understand why sustainability will not arrive on our doorsteps like a package. It will not be found in a particular plan or product. Rather, it can be experienced only as action grounded in awareness and as a new way of seeing that arises from a new way of doing.

In Antoine de Saint-Exupéry's *The Little Prince*, the stranded pilot is asked by the boy to draw a sheep. None of the attempted sketches pleases the little prince. Finally, the pilot draws a parallelogram, announcing it as the box in which the sheep is sleeping. The little prince is delighted. Like the artistically challenged pilot, I cannot produce a detailed picture of a sustainable future replete with green guidelines and gadgets. Instead, I offer an account of the kind of thinking and behavior that got us into our current predicament. In turn, and more important, I explore the sensibilities and practices required to chart a course to more hopeful seas.

The Habitats of Contemporary Life

An *oikos* is a habitat or dwelling place, a spatial realm characterized by a network of relationships. In this book, I expand the meaning of *oikos* to include the many distinct, albeit interconnected, networks of relationships that we inhabit today. Our habitats, in this respect, are not limited to geographical locales. They may include any system of interconnected associations characterized by identifiable actors, dynamic patterns of interaction, and established, if ever transforming, meanings. An *oikos* is a nested realm of complex interdependence.

We often speak of the "world of art" or the "realm of law." We are referring to a system of interconnected associations—aesthetic or legal in this case—characterized by specific actors, modes of interaction, and meanings. This book examines eight such habitats, eight swatches of the fabric of life. They are the distinct yet interconnected domains of ecology, ethics, technology, economics, politics, psychology, physics, and metaphysics. These fields of inquiry and facets of life do not exhaust the contemporary land-scape. I leave out the aesthetic and legal realms, as well as the disciplines of medicine, chemistry, mathematics, communications and media, and many others. But the habitats addressed in the following chapters most powerfully display the growing interdependencies of contemporary life.

The first chapter investigates ecology. Ecology pertains to the study of ecosystems, the networks of relations maintained and transformed by populations of diverse species in the common pursuit of sustenance, physi-ological growth, and reproduction. There are many distinct ecosystems on the planet, though with the possible exception of deep-sea vents, remote islands, or oases of life surrounded by impassable physical barriers of sand or stone, such communities of life are never wholly self-enclosed. Increas-ingly, the planet's ecosystems are interconnected by global phenomena such as climate change and the transportation and dissemination of pol-lution and species. Ultimately ecology is the study of the biosphere, the combination of the planet's ecosystems. The realm of ecology is the dwell-ing place of biological life. Safeguarding this *oikos* is the central task of sustainability.

Chapter 2 explores the realm of ethics—the network of values and norms we establish with our fellow men and women, and, potentially, the values and norms we assume regarding other species. These may take the form of abstract principles and standardized rules of social conduct or more diffuse regimes of comportment—often labeled virtues—that contribute to the "good life." Ethics is a place of rights and responsibilities, reciprocity and obligation, character development and personal conduct. At base, ethics refers to the principled practices we develop to sustain the commu-nities that sustain us. To exist in the moral realm is to be occupied with the rules, modes of behavior, and understandings that facilitate well-governed, just, and beneficial communities. Sustainability is an ethical sensibility and commitment—arguably the ethical sensibility and commit-ment most in need of cultivation today.

Technology is the topic of chapter 3. Technology pertains to tool-making, machine-building, and, in its most advanced form, the crafting of artificial forms of intelligence and life. While the technology of our prehistoric forebears was very limited and relatively sparse—a few wooden clubs and spears, a flint for making fire, a stone axe—the world of contem-porary humans is largely defined by tools, machines, and other engineered processes and products. Indeed, it is difficult to imagine our lives apart from the technological capacities and artifacts that we develop and deploy at an accelerating rate. Technology has made its indelible mark on the world and on the human species itself. It increases our ability to achieve concrete goals with ever-greater efficiency while simultaneously heightening the peril of unintended consequences. Characterized by the manipulation and mastery of the world through innovation and craft, technology both threatens the sustainability of civilization and defines its development.

Chapter 4 focuses on economics, the world of market relationships. The economic realm is a place of production and exchange—the making, buying, selling, and bartering of goods and services. Notwithstanding the dictum of caveat emptor, that the buyer must beware, relationships within the economic realm typically rely on social trust. This trust is grounded in common practices, in the rules of fair play, and in laws established by political regimes to govern market transactions. Notwithstanding such ethical and political foundations and the affective bonds, norms, and cooperative pursuits that develop through them, economic relations are driven by the engine of self-interest. The way we collectively organize the pursuit of self-interest bears directly on the public benefits and ills that economic life produces. Economic pursuits divorced from ecosophic awareness threaten the very fabric of life today.

The arena of politics occupies us in chapter 5. Politics is the place where power is publicly generated and used to define and pursue public goods. A crucial public good is a healthy environment. Another public good within democracies is power itself. To be sustainable, a democratic political system must foster an equitable sharing of power as a public good and a means to the equitable sharing of other goods (as well as responsibilities and risks). The relationship of democratic politics to sustainability is one of the most intriguing and important topics of concern today. Understanding the fundamental interdependencies of political life, and the meaning of freedom within these public relationships, may inspire us to conceive anew the challenge of sustainability.

Chapter 6 addresses psychology—the realm of the mind, self, or soul. The *psyche*, which is what the ancient Greeks called the soul, is an inner dwelling place, and like any other habitat, it is a place of relationships. As self-conscious creatures, human beings establish relationships with themselves. But the self is not a simple dyad. Each of us, as Walt Whitman famously stated, is a multitude. Psychological health depends on the proper development of the soul's distinct parts and their proper integration. Psychological health is related to ecological and social health. What we do in and to the world outside largely depends on the state of our inner worlds. Our external relationships reflect and have an impact on the network of connections within. Sustaining the world and sustaining our souls are synergistic endeavors.

The fields of physics and metaphysics are combined in chapter 8. Physics and metaphysics explore and situate us within our cosmic dwelling place. Physics pertains to the empirical world of matter, energy, and their patterns of generation and transformation. Metaphysics, though it does not reckon with empirical observation or experiment, is equally concerned with universal laws and relationships. Eternal questions, such as the ultimate nature of being, are its mainstay. Contemporary physics explains normal cause-and-effect relationships in our universe but also finds evidence of more encompassing and pervasive forms of interdependence. At a quantum level, physics presents us with seemingly nonmaterial forces at play. Experimental evidence today suggests a level of connectedness in the cosmos that ancient metaphysicians first hypothesized. It is through the realms of physics and metaphysics that we glimpse the full breadth and depth of interdependence and the cosmological context of unintended consequences.

A few decades ago, it was fair to say that an ecological "view of existence" and an appreciation of interdependence were "alien to Western ways of looking at things."[7] Today it infuses a wide array of our disciplines and practices. Yet these domains of knowledge and practice, historically and now, often acknowledge specific aspects or features of interdependence only to disregard its pervasive presence and profound implications. The chapters that follow highlight the growing recognition of interdependence within eight fields of inquiry and facets of life. They also address the tendencies within these habitats to ignore the full depth and breadth of interdependence and its implications.

It may be obvious why and how ecology, ethics, technology, economics, and politics find their place in a book dedicated to better understanding and caring for the web of life. These topics are standard fare in research and writing concerned with sustainability. But why would readers interested in sustainability—people rightfully concerned about climate change, worried about overpopulation and resource depletion, pained by the extinction of species, intrigued by the opportunities for renewable energy, and committed to social empowerment in a divided world—grapple with chapters devoted to psychology, physics, and metaphysics? What does sustainability have to do with the structure of the human psyche and the composition of the cosmos? The answer to this question highlights the radical claim of *Indra's Net and the Midas Touch*.

Today *sustainability* is dangerously close to becoming a moniker for the engineering of a particular sort of world. Yet it is precisely humanity's increasing predilection for world making, and our increasing power to do so, that threatens its sustainability. To become at home in a self and a world never fully of our making or within our control is the core challenge of sustainability. Meeting this challenge takes us on a voyage into a mysterious universe and our multifarious souls.

Indra's Net

In the early Pali scriptures of India, one finds the doctrine of *paticca samuppada,* which means dependent co-arising or interconnected origination. The notion is that all physical and mental phenomena come into existence and develop as interdependent relationships. Nāgārjuna (c. 150–250 C.E.), the founder of Mahayana Buddhism, denied the existence of isolated entities bearing essential natures. He argued that all beings were "empty" of separate, distinct essences. Within Mahayana Buddhism, the Avatamsaka tradition developed Nāgārjuna's notion of śūnyatā or fundamental emptiness along with the understanding that relationships of interdependence were the stuff of which the universe was made. The common tendency to perceive the world in terms of independent entities was identified as the cause of suffering (*duhkha*).

In the sixth century, the Hua-yen school of Chinese Buddhism (which flourished in the T'ang dynasty) addressed the nature of interconnectedness. Hua-yen was a philosophically oriented school that explored the metaphysics of the (meditative) practices of Ch'an Buddhism (Zen in Japan). It took its inspiration from the Avatamsaka sutra or Flower Garland scripture, originally written in Sanskrit but translated and completed in Chinese. The sutra described the universe as "one great scheme of interdependency."[8] A favorite topic of the Hua-yen school is the story of Indra's net.

Indra is the lord of heaven and the king of the Vedic deities. Over his palace on Mount Meru, the *axis mundi* of Vedic cosmology, hangs a net that stretches infinitely in all directions. At each node of the net, where the heavenly strands intersect, hangs a jewel. Stretching across the unending breadth of the universe, the jewels are infinite in number. They are also infinite in composition. Each facet of each jewel reflects all the other jewels hanging from the net.

The brilliant jewels presenting an infinite cavalcade of reflections are stunning, but they have no independent essence. The jewels of Indra's net are not enduring substances. Rather, each jewel is manifested only as a reflection of all the other jewels. Each gem owes its existence to the network of reflections to which it contributes.[9]

The story of Indra's net is meant to illustrate the interdependence and interpenetration of phenomena. All the strands of the net are connected. Loosen one, and all are loosened. Sever one, and the whole is weakened. This is the meaning of interdependence. Like the filaments of Indra's cosmic net, the jewels hanging from its vertices are all interconnected. In turn, they also mirror each other. Indeed, they are constituted—brought into reality, as it were—through this reflective relationship. Here the part is not only connected to the whole by way of multiple linkages. The part actually includes the whole. Each jewel contains (the reflections of) all the other jewels and is, in turn, contained by them. This is the meaning of interpenetration.

Interpenetration might be thought of as an intensified, deepened form of interdependence. Interdependence refers to things existing in connection. Interpenetration asserts that connectedness itself (rather than things existing in connection) constitutes the most fundamental reality. When we focus on connections rather than things, we discover that the parts (the things connected) reflect and sustain the whole (the network of relationships) as much as the whole reflects and sustains its parts.

The tale of the jewel net of Indra strikingly foreshadows contemporary ecological thought, which is equally focused on relationships of interdependence rather than isolated organisms or entities. It presents us with a "cosmic ecology."[10] But this ancient tale of connectivity prefigures more than ecological discourse. The chapters of this book explore fields of inquiry whose subject matters are increasingly grounded in an appreciation and awareness of the pervasiveness of interdependence. *Indra's Net and the Midas Touch* provides evidence for and understanding of the heightened connectedness within and across distinct realms of contemporary life, with the intent of broadening and deepening the burgeoning prospects and practices of sustainability.

When portrayed as the required remedy for our dire straits, as the only alternative to catastrophe, sustainability can be a bitter pill. It portends sacrifice. To be sure, becoming attuned to the interdependence and interpenetration of all things fosters restraint and prudence. But it also stimulates creativity and community. The manifold relationships of which the fabric of life is woven need not immobilize us. Our attentive participation in them allows us to travel the path of sustainability with hope. Awareness of the breadth and depth of our connectedness is a profound responsibility, and a blessing beyond measure.

1 Ecology

When you try to pick out any thing by itself, you find it hitched to everything else in the universe.

—John Muir

Our actions not only have effects; they also have side effects, which are generally unanticipated, unpredictable, and all too often pernicious. In 1963, an ecologist and microbiologist from the University of California in Santa Barbara coined a phrase to capture this unforgiving phenomenon: "We can never do merely one thing."[1] Its author, Garrett Hardin, believed his words to be "splendidly original." More research would have him acknowledge that the insight, dubbed the first law of human ecology, could be found within the wisdom traditions of numerous cultures.[2] Pithy aphorisms from sages well convey the insight. Illustrative examples are also to be found in ancient fables and myths that testify to the unanticipated, and often disastrous, effects of human effort.

The tale of King Midas, recounted by Ovid and other bards, captures the gist of Hardin's law. Midas, the king of Pessinus, is walking in his rose garden when he happens on Silenus, a satyr, who had fallen asleep drunk. The king treated the trespasser with kindness. On learning of Midas's benevolence, the god Dionysus, who was often accompanied and tutored by the satyr, granted the king one wish. Midas, more than a little fond of wealth, requested that everything he touched be turned to gold. The wish was duly granted, and Midas gleefully went about experimenting with his new powers. He rejoiced as he wandered the countryside, transforming stones and trees along his path into the purest gold.

Commanding his servants to set before him a grand feast to celebrate his new talent and riches, Midas quickly discovered the unintended consequences of his alchemy: all food and drink hardened to inedible metal as it touched his lips. Faced with certain starvation in the midst of plenty, Midas begged Dionysus to rescind his wish. The god complied, and the king once again thrived, a little poorer but well fed and much the wiser.

The basic truth behind Hardin's succinct phrase and Ovid's entertaining tale received its first social scientific analysis in 1936 by Robert K. Merton, a prominent sociologist at Columbia University, who coined the term "unanticipated consequences." In a journal article devoted to the topic, he observed that a systematic and scientific investigation of the problem of the unanticipated effects of "purposive" action was long overdue.[3] Purposive action refers to deeds deliberately done for explicit purposes. It is action directed toward the achievement of goals. The law of unintended consequences, as the phenomenon came to be known, dictates that whatever one's motives and however successful one's efforts to attain specific outcomes, other outcomes that are neither intended nor anticipated inevitably will follow.[4]

Merton suggested that our actions produce unintended and often unwanted consequences owing to a number of factors. These include our reliance on habitual behavior in the face of changing circumstances, errors in calculation, and beliefs, values, or prejudices that obscure our (long-term) interest. But unintended consequences also arise from simple ignorance. The "interplay of forces and circumstances," Merton observed, are often "complex and numerous."[5] It is practically impossible to calculate them all. Over the next half-century, Merton drafted more than six hundred pages of manuscript in an effort to come to grips with the issue, but died before completing the project.[6]

Writing at the time of the New Deal, a period of unprecedented expansion for the federal government, Merton was primarily concerned with the unintended consequences of social welfare programs and economic planning. Hardin, sharing similar worries about the unintended consequences of social reform, included environmental repercussions in his purview. Best known for his 1968 essay "The Tragedy of the Commons" (originally published in *Science* but since reprinted in over one hundred anthologies), Hardin links his insight regarding the fickleness of human action to the fields of economics, politics, ethics, and population theory. Its impact in the ecological realm gained the most attention.

The law of unintended consequences has become a core principle of contemporary environmental thought. Lester Milbrath claims that it constitutes the "central axiom of environmentalism." Knowing that we can never do just one thing, Milbrath observes, "we must learn to ask, for every action, And then what?"[7] It is a question too seldom asked. Consequently, environmental scientists, no less than environmental activists, find their careers defined by the need to remedy the unintended ecological effects of human enterprise. One such scientist and activist was Rachel Carson.

Rachel Carson and the Rise of Environmentalism

No species lasts forever. Today, however, the planet's plants and animals are going extinct a thousand times faster than would occur if humankind were not among them. And the evolution of many new species, primarily vertebrates, has ground to a halt for the same reason. The most comprehensive inventory of the status of the earth's plants and animals, compiled by the International Union for the Conservation of Nature, lists 17,000 species as threatened with extinction, including one-fifth of known mammals, almost a third of all amphibians, over a third of freshwater fish, and more than one out of ten of the world's birds.[8] The five major causes of this tragic loss of biodiversity are habitat loss, climate change, invasive species, pollution, and predation. Only one of these causes—predation or overexploitation through hunting, gathering or fishing—might be considered a conscious effort to diminish a species. All of the other major causes produce the extinction of plants and animals as an unintended consequence of human activity directed to other ends. Even in the case of predation, the extinction of a species is hardly an intended consequence. Indeed, for those who make their living by harvesting wild species, such as fishermen, the decline or extinction of this resource is an unmitigated disaster.

Rachel Carson's *Silent Spring*, published in 1962, provided the general public for the first time with a stunning account of the loss of biodiversity owing to the unintended consequences of human action. With its publication, Carson jump-started the contemporary environmental movement. For this and other reasons, *Silent Spring* has been dubbed one of the twenty-five greatest science books of all time.[9] It made the law of unintended consequences—without ever mentioning the phrase—a household concern. Hardin considered Carson's classic study a "monument" to the insight that you can never do just one thing.[10]

Before penning *Silent Spring,* Carson had already made a name for herself as a writer of natural history. Trained as a marine biologist, her earlier books, *Under the Sea Wind, The Sea around Us*, which won a National Book Award and topped the national best-seller list, and *The Edge of the Sea,* highlighted the wondrous interdependencies among species in marine habitats. Carson gave frequent public lectures on this topic, including one presented to the American Association for the Advancement of Science in 1953. Later that decade, Carson's reputation as a nature writer would thrust her into the maelstrom of history. But it was the denizens of the sky, not the sea, that now captured her attention.

In 1958, having discovered many dead birds in her backyard birdbath after an aerial spraying of DDT for mosquitoes, New England resident Olga Owens Huckins published a letter in the *Boston Herald* addressing the issue. She sent a copy of the letter to Carson, whose writings in natural history she admired. Huckins's letter pushed Carson to move pesticide use—which she personally had been concerned with for over a decade—to the front of her professional agenda. In May 1958, Carson signed a contract with Houghton Mifflin to write what she thought of as "the poison book," tentatively titled, *Man against the Earth*. That summer, Carson met with the editor of the *New Yorker*, William Shawn, to sign a contract for a series of articles on the potential dangers that pesticides posed to wildlife and human health. The *New Yorker* articles appeared in June 1962, and Houghton Mifflin put their compilation in book form on store shelves in September.

The message of *Silent Spring* was as straightforward as it was ominous. Pesticides did not just kill noxious insects. They had the unexpected effect of decimating entire populations of other animals, including many birds. Absent its beloved birds, American neighborhoods would face a silent spring, not only free of pesky bugs but barren of avian singers. Carson explained how the age-old attempt to gain "control of nature" was self-defeating because it failed to comprehend, or preserve, the intricate relationships that constitute "the whole fabric of life."[11]

The key weapons employed in the pursuit of the mastery of nature were chemicals. Dichloro-diphenyl-trichloroethane (DDT) was developed in

1939 by chemist Paul Mueller, who won the Nobel Prize in 1948 for inventing this poison against arthropods. DDT was employed widely in the South Pacific during World War II to control the spread of malaria among U.S. infantry and as a delousing powder for troops in Europe. The chemical saw a slow introduction to civilian use after 1945. By the mid-1950s, the U.S. Department of Agriculture (USDA) had initiated a multistate campaign employing DDT in aerial sprayings to control the spread of the gypsy moth, which was devastating the deciduous forests of the Northeast. The pesticide was also used to eradicate other insects, and it was not long before the unintended consequences of this effort to control nature with industrial chemicals showed up in Olga Huckins's backyard.

Before the third installment of Carson's series of articles in the *New Yorker* appeared, chemical companies threatened her with lawsuits, and McCarthyesque accusations were leveled. Critics suggested that heeding Carson's warnings about the dangers of chemicals would reduce America's agricultural productivity to "east-curtain parity."[12] Indeed, Ezra Taft Benson, President Eisenhower's secretary of agriculture, concluded that Carson was "probably a communist," adding that he did not understand "why a spinster with no children was so concerned about genetics."[13] The ad hominem attacks, as painful as they might have been to the genial and soft-spoken Carson, heightened the publicity for her work. *Silent Spring* became a national best-seller, with hundreds of thousands of copies sold within months of its publication. It was selected by the Book-of-the-Month Club and received an endorsement from Supreme Court Justice William O. Douglas, who argued in favor of a "Bill of Rights against the 20th century poisoners of the human race."[14] President Kennedy would acknowledge that Carson's work stimulated the USDA's eventual investigation into the danger of pesticides. In 1963, Carson met with the President's Science Advisory Committee to address her concerns. This meeting began a process of citizen advocacy and government response that culminated in banning DDT in the United States in 1972.

Like many other pesticides, DDT proves ecologically devastating because its application cannot be isolated to targeted species. All biological forms of life, plant and animal, exist as strands of an ecological web. Many of these interconnected relationships are symbiotic, meaning they are beneficial to both participants. Others are competitive or predatory. In competitive relationships, different species vie for the same food source or habitat.

In predatory relationships, one species makes another its source of food. The pernicious side effects of DDT arise because of the role it plays in such food chains.

Along with many other pesticides and certain heavy metals, DDT is prone to bioaccumulation and biomagnification. The pesticide is not expelled from the body once absorbed. Rather, it accumulates in tissues, typically in fat deposits. Species higher on a food chain suffer an increasing level of contamination because they incorporate the accumulated toxins from each of their food sources. Whereas DDT may be found in negligible quantities in the waters of treated areas, for example, it will appear in greater concentrations in the plankton, in still greater concentrations in the small fish that eat the plankton, and in even higher concentrations in the large fish that eat the small fish. It will be at its greatest levels in the raptors that prey on the large fish. Owing to this biomagnification, ospreys were found to contain levels of DDT many million times greater than that of the water over which they fished.

Indeed, it was an effort to protect ospreys—whose thinning eggs from DDT exposure caused Long Island populations of the birds to plummet—that led to the formation of the Environmental Defense Fund in 1967 and the outlawing of the pesticide five years later. Although DDT was banned in the United States, other countries that provide the wintering grounds for migratory birds continue to use the poison, along with other organophosphate pesticides. As a consequence, many North American bird populations continue to decline.[15]

The onset of the contemporary environmental movement is often attributed to *Silent Spring*'s vivid description of the unintended consequences of pesticide use. Carson did not argue for the outlawing of all pesticides, which she suggested should really be called *bio*cides owing to their effect of destroying multiple forms of life. Rather, she insisted that the "indiscriminate" use of biocides was too dangerous to tolerate. When it comes to the problem of insect pests, she argued, the chemical cure often proves worse than the natural disease.

DDT saw its first widespread civilian application in aerial sprayings to combat the "disease" of the gypsy moth. Ironically the gypsy moth itself was the unintended consequence of an earlier, equally well-intentioned action. The moth is not native to North America. In the late 1860s, it was accidentally introduced by an amateur entomologist fascinated by silkworms. After a return visit to his native France, Etienne Leopold Trouvelot brought back to his working-class suburb of Boston a clutch of gypsy moth eggs. His plan was to breed and study the insect. Unfortunately some of the larvae escaped the backyard laboratory. Trouvelot informed the authorities of the mishap, but no action was taken. Eventually Trouvelot lost interest in his experiments with insects. He took up astronomy instead, a field in which he would prove much more successful, eventually teaching at Harvard University and having a crater on the moon named in his honor. But the gypsy moths did not die along with Trouvelot's interest in them. His abandoned hobby would produce a plague.

By the time Trouvelot returned to France in 1882, his former Boston neighborhood was infested with the moths. Particularly fond of the grand oak trees that provided streets with shade, scenic beauty, and habitat for birds, the insects did not stop at the end of Trouvelot's block. The whole city became bug-ridden. Indeed, the moths did not limit themselves to urban life. By 1889, gypsy moths were destroying increasing acreages of the state's forests, and the Massachusetts State Board of Agriculture became involved. Its efforts to burn infected trees, manually remove egg masses, and apply primitive insecticides proved ineffective at controlling the insects. Within a decade, the Massachusetts program to eradicate the gypsy moth was deemed a lost cause and terminated.

Although other campaigns were mounted in the region, including a large-scale effort by the USDA in 1957, the gypsy moth continued to spread (mostly by wind, it so happens, because females are incapable of flight). The moth's current range now includes the entire northeastern United States, as well as portions of the Southeast and Midwest. Its domain continues to expand by over 20 square kilometers a year.

In innocently pursuing his experiments, Trouvelot inadvertently introduced a plague. And in its effort to eradicate the disease that the amateur entomologist created, the government produced an equally pernicious, unintended effect. We are thus presented with a fatal irony: the deadly plight that Rachel Carson identified was the consequence of a cure. This was neither the first nor the last time that an ingenious solution to a stubborn challenge proved itself a more wicked problem.

Thinking Like a Mountain

If Rachel Carson deserves credit for thrusting ecology into the national limelight, Aldo Leopold must be honored for setting the stage. Leopold's *A Sand County Almanac*, now famous, was published to little acclaim in 1949, a year after his death. Sales of the book on its release were slow and unimpressive; only a few thousand copies were purchased over the following decade. But eventually the magnitude of Leopold's achievement became apparent. As one environmental historian states, *A Sand County Almanac* "signaled the arrival of the Age of Ecology."[16]

In 1909, Leopold obtained his master's degree from the Yale School of Forestry, which had been endowed by the family of Gifford Pinchot, America's leading forester and the father of conservation. Leopold then worked under Pinchot in the Forest Service, serving for the better part of two decades. The agency had come to control increasing acreages of public land established as national forests under the 1891 Forest Reserve Act. Leopold's boss was an advocate of the "wise use" of these public lands. Schooled in German and French forestry practices, Pinchot advocated the scientific management of natural resources. He pitted himself against the "boomers" and "land grabbers" of his day—men who recklessly plundered western lands for their mineral wealth and timber.

Outlining his philosophy of conservation, which might stand today as a passable definition of sustainability, Pinchot wrote: "The central thing for which Conservation stands is to make this country the best possible place to live in, both for us and for our descendants. It stands against the waste of natural resources which cannot be renewed, such as coal and iron; it stands for the perpetuation of the resources which can be renewed, such as the food-producing soils and the forests; and most of all it stands for an equal opportunity for every American citizen to get his fair share of benefit from these resources, both now and hereafter."[17] Leopold adopted Pinchot's philosophy of conservation, which entailed the active management of a nation's natural resources, including wildlife, to maximize the long-term benefits for citizens. Then, quite abruptly, Leopold's philosophy of conservation was transformed, as was his worldview, after an encounter with wolves.

In 1914, a federal predator extirpation campaign was initiated on all public lands. Under the presumption that predators deprived deer and elk hunters of their prey and were the chief threat to ranchers' livestock, wolves, mountain lions, and coyotes were systematically shot and poisoned. The extermination program was combined with an effort to increase the deer population by clearing forests so that meadows would form where deer might graze. As a public servant in charge of national forests in New Mexico and Arizona, Leopold was party to these policies. He eagerly participated in efforts to "clean out" wolves, mountain lions, and coyotes. Whenever possible, he shot predators and directed others to do the same.

The program was a smashing success: wolves were virtually eliminated in the contiguous United States by 1935, livestock kills were down, and deer, lacking natural predators to winnow their herds, vastly increased in numbers. The amount of land they depended on for forage, however, did not increase, and in short order, the deer overshot the carrying capacity of their habitat. At this point, the options were grim: either the deer had to be shot in droves to bring their populations back in balance with the carrying capacity of the land, or they faced mass starvation and disease from their weakened states. In either case, the formerly bountiful species would soon be, in Leopold's words, "dead of its own too-much." Such drastic fluctuations were not good for the ungulates. The land did not fare any better. As Leopold wrote:

I have lived to see state after state extirpate wolves. I have watched the face of many a new wolfless mountain, and seen the south-facing slopes wrinkle with a maze of new deer trails. I have seen every edible bush and seedling browsed, first to anemic desuetude, and then to death. I have seen every edible tree defoliated to the height of a saddle horn. Such a mountain looks as if someone had given God a new pruning shears, and forbidden Him all other exercise. In the end the starved bones of the hoped-for deer herd, dead of its own too-much, bleach with the bones of the dead sage, or molder under the high-lined junipers.[18]

Leopold came to recognize that the human management of nature had inherent limitations. Although Pinchotian conservation benefited from an expanded time horizon and laudable democratic values—rightly standing opposed to the shortsighted self-interest of boomers and land grabbers—it was not yet ecosophic. Although it aimed at long-term benefits for the greatest number of human resource users, it did not well understand the interdependent web of life that it purportedly managed. As a result, it failed to comprehend that land managers can never do just one thing. Excising one strand of the ecological web—killing predators—did not lead to a flourishing of the predators' ecological partners. Rather, it led to the rapid unraveling of the web of relations on which both predators and prey depended. To avoid such devastations, Leopold counseled, we must change our role from "conqueror of the land-community to plain member and citizen of it."[19]

The notion of ecological citizenship developed slowly for Leopold in the face of denuded mountain slopes and deer bones. Even with the evidence before him, the public employee continued to carry out his mandate of predator extirpation. Then came Leopold's epiphany. The full meaning of membership in a land community occurred in the midst of a violent effort to gain control of it. Leopold recounts the episode:

We were eating lunch on a high rimrock, at the foot of which a turbulent river elbowed its way. We saw what we thought was a doe fording the torrent, her breast awash in white water. When she climbed the bank toward us and shook out her tail, we realized our error: it was a wolf. A half-dozen others, evidently grown pups, sprang from the willows and all joined in a welcoming melee of wagging tails and playful maulings. What was literally a pile of wolves writhed and tumbled in the center of an open flat at the foot of our rimrock. In those days we had never heard of passing up a chance to kill a wolf. In a second we were pumping lead into the pack, but with more excitement than accuracy: how to aim a steep downhill shot is always confusing. When our rifles were empty, the old wolf was down, and a pup was dragging a leg into impassable slide-rocks. We reached the old wolf in time to watch a fierce green fire dying in her eyes. I realized then, and have known ever since, that there was something new to me in those eyes—something known only to her and to the mountain. I was young then, and full of trigger-itch; I thought that because fewer wolves meant more deer, that no wolves would mean hunters' paradise. But after seeing the green fire die, I sensed that neither the wolf nor the mountain agreed with such a view.[20]

Leopold's encounter with the dying wolf was transformational, prompting the development of an ecological perspective he dubbed "thinking like a mountain." To think like a mountain required a full appreciation of the vast and intricate web of interdependent relationships that constituted a mountain *oikos*. Thinking like a mountain is an exercise in ecosophic awareness.

To act as a citizen of a mountain habitat rather than its conqueror does not mean that one turns the mountain into a museum and becomes its curator. To the extent that we participate as plain members of a natural community, our interactions will inevitably entail consumption and use.

Becoming part of a food chain, after all, is essential to ecological citizenship. Owing to our growing appetites and numbers, however, the human use and consumption of other species demands a form of management mindful of the fabric of life. As managers, we may tinker with the land community. But this tinkering, heeding the inevitability of unintended consequences, must be a humble affair. Extermination of a co-member of the community is forbidden. As Leopold wrote, "The last word in ignorance is the man who says of an animal or plant: 'What good is it?' . . . If the biota, in the course of eons, has built something we like but do not understand, then who but a fool would discard seemingly useless parts. To keep every cog and wheel is the first precaution of intelligent tinkering."[21]

Ecological citizenship does not demand the impossible task that we remove ourselves from the web of life. Indeed, greater integration is often required. But the project of conquest must be abandoned. The lesson that Rachel Carson attempted to teach in a book originally entitled *Man against the Earth* still stands. In the prescient words of Morris Berman, "If you fight the ecology of a system, you lose—especially when you 'win.'"[22]

A Cascade of Unintended Consequences

All tinkering with ecosystems produces unintended consequences. Unintelligent tinkering can have drastic results. A common outfall is the cascade effect. Cascade effects, also known as trophic cascades, occur within ecosystems defined by strong (symbiotic, competitive, or predatory) interactions where one species has been removed or its relation to other species has been drastically altered. A well-documented example occurred as the result of overharvesting sea otters in the Pacific Northwest. The simple act of trapping a wild animal for commercial purposes (the otters were prized for their thick pelts) led to the wholesale transformation of marine communities.

Otters prey on sea urchin, one of their favorite foods. As otter populations declined drastically in the nineteenth century with increased trapping, the urchin populations on the Pacific coast exploded. Urchins have their own favorite food: kelp. Rising urchin populations led to the overgrazing of kelp, creating "urchin barrens," the oceanographic equivalent of the deer-denuded mountains that Leopold described.[23] Kelp forests support a broad variety of fish and other species, and are the breeding grounds for a large share of invertebrate sea life. The overhunting of sea otters eventually led to the local extinction of many other species, radically changing the composition of the aquatic community.

Nature abhors a vacuum. When the urchins decimated the kelp forests, the coastal habitat that the kelp formerly occupied was open to invasion by Asian macroalga and European bryozoan, two forms of algae. As a result, the kelp forests, and the communities of life they supported, often failed to rebound even when the overharvesting of otters ceased. A cascade effect can fundamentally transform an entire ecosystem. Once unintelligently tinkered with, communities of life may never recover.

Examples abound of cascade effects, aquatic and terrestrial, that began with the overharvesting of a particular species and ended with the devastation of ecosystems. What typically varies is the number of links in the disrupted food chain. Consider the same cascade effect with a few extra links added. In Alaska's coastal waters, the overfishing of perch and herring by factory trawlers depressed the population of sea lions and seals, which fed on the once abundant fish. The decline in sea lions and seals forced the resident orcas, often known as killer whales, to feed on otters rather than their preferred (larger and more blubbery) prey. From here the story takes a familiar turn. With otters in short supply, growing urchin populations destroyed kelp forests and the species that depend on them.

Not every change within an ecosystem produces a cascade effect. Contrary to early environmental thought, ecological communities are not inert. Indeed, the equilibrium that they demonstrate is gained through persistent fluctuations. Nature is never static. Talk of the "balance of nature" is misleading in this regard. Ecological communities are inherently erratic, variably respond to disruption, and have characteristics that shift with the tides of time. Nature is incessantly engaged in its own tinkering. Frequent small-scale changes facilitate long-term stability. This is known as resilience.[24] Ecosystems become resilient not because they forbid change, but because they respond well to it. They adapt. As in all other things, however, moderation proves key. When disruptions of patterns and relationships between species are too pervasive, too intensive, or too abrupt, an ecosystem may prove incapable of adapting in time, and it collapses. Human-induced disruptions frequently occur at a scale and speed that overwhelm an ecosystem's capacity to respond adaptively.

Disruptions need not be the result of harvesting that gets out of hand, as was the case in aquatic communities where the fishing of perch and herring or the trapping of otters led to collapse. Often devastation is the unintended consequence of human activity that is not at all predatory. The inadvertent introduction of exotic species to established ecosystems is a case in point. Trouvelot's unsuccessful experimentation with gypsy moths is just one of many troubling examples that span the globe. Like Trouvelot, the people who introduce species into new environments typically intend to manage and control their exotic acquisitions. Alternatively they may assume that the new species, being a child of nature itself, will simply blend into the local biological family. But an important contribution to the web of life in one ecosystem may become an unstoppable bane in another, where it finds no natural predators or barriers to expansion. A few of the more infamous cases of invasive species will illustrate the point.

In the mid-1930s, after learning of Hawaii's success in battling sugar-cane-eating beetles with cane toads (*Bufo marinus*), Australia introduced thousands of the amphibians into its cane fields. Unfortunately Australia's sweet-toothed beetles can fly, unlike the Hawaiian variety. As a result, the Australian beetles typically chow down at the top of cane stalks. Cane toads, indigenous to Central and South American where there is plenty of food to be found at ground level, are not very good jumpers. So while they feasted on Hawaii's crawling beetles, they were flummoxed by Australia's winged variety. Before the end of the decade, officials announced that the cane toad experiment had failed.

But the amphibians did not starve. Unable to catch flying beetles, the toads turned their sticky tongues on other prey, and they found plenty to their liking. They also appreciated the absence of natural predators in their adopted land. With a relatively quick development period and females capable of producing over 30,000 eggs each year, the population of cane toads exploded, eventually reaching 200 million and spanning an area many hundreds of thousands of square miles. In infested areas, animals that fed on the amphibians declined precipitously, because the toads have toxic skin. Marsupials and reptiles, including rare crocodiles, suffered this fate. Native frogs whose habitats were colonized by the toads were decimated. In turn, populations of seed-harvesting ants and other insects that became a key part of the cane toads' diet were wiped out, causing a disruption in the area's vegetation dynamics.

The Australian government has spent millions of dollars in efforts to eliminate, or at least limit, the expansion of the toxic toads. Efforts have met with little success, and experiments with genetic engineering are now in development. The cane toads have achieved such notoriety on the southern continent that football teams have been named after them. More vengefully, the ubiquitous amphibians have become the unsanctioned, but not infrequent, objects of abuse in games of cane toad cricket. Batting a "sticky" wicket has taken on a whole new meaning in Australia.

The most pernicious foreign species are often not unsavory, predatory animals. In many cases, the worst of the invasive exotics are lovely green plants. Kudzu is an infamous example. A climbing vine that grows quickly, kudzu creates a thick mat of broad, dark-green leaves that covers everything in its path, effectively crowding out other flora on which fauna might feed. Native to Japan and China, the kudzu plant was brought to the United States in 1876 for the Philadelphia Centennial Exhibit, just about the time that Trouvelot's moths infested Boston's streets. Introduced to Florida in the following years, kudzu was avidly promoted as a forage crop, for decorative purposes, and for erosion control. As it turns out, kudzu liked its new abode in Florida and grew even fonder of neighboring states. Consequently it is now none-too-affectionately known as "the plant that ate Georgia." Locals are wont to photograph what used to be fences, cars, or entire buildings now completely buried under a thick canopy of kudzu. Currently the plant is listed as invasive in twenty-two states, covering between 20,000 and 30,000 square kilometers of land. The cost of controlling the spread of kudzu and mitigating its damage runs over half a billion dollars a year.

What kudzu achieves on land is accomplished in lakes and rivers by the water hyacinth. Native to the Amazon basin, water hyacinth was introduced to North America in 1884 as decorative foliage to a backyard pond. It has since gained the reputation as the worst aquatic weed in the world, blocking waterways; depriving ponds, lakes, and streams of sunlight and oxygen; crowding out native species; and producing fish kills in more than fifty countries across five continents. But invasive species are not always experiments in transplantation that get out of hand. The introduction of foreign species to a new habitat is often the unintended consequence of a seemingly unrelated human activity. Such was the case with the construction of the Erie Canal and the transporting of old guns and mortars to the island of Guam.

The Erie Canal, built between 1817 and 1825, connects the Hudson River in New York State to Lake Erie, effectively joining the Great Lakes to the Atlantic Ocean. More than 360 miles long and mostly dug by hand, the canal was an engineering marvel of the century. Its construction can be credited with spurring the westward expansion of settlers and made New York the nation's premier commercial city. The canal is also credited with introducing the sea lamprey to the Great Lakes, allowing these fish to bypass the natural barrier of Niagara Falls.

Sea lampreys are eel-like parasites that attach themselves to fish with their suction cup mouths and razor-edged teeth. They voraciously feed off the bodily fluids of their hosts, weakening or killing them in the process. Between the introduction of the sea lamprey to the Great Lakes in the mid-1800s and the 1960s, many native fish became rare or extinct, and the once-thriving fishing industry collapsed. The catch of lake trout dropped to one-fiftieth of its former levels. Officials are still hard at work after more than a half-century to control the parasitic fish.

The sea lamprey is not alone in its status as a noxious Great Lakes exotic. As a result of the building of the Erie and other canals, over 150 invasive species have been documented in these waters. The introduction of the lamprey, however, was doubly dangerous: first in its direct role as an invasive exotic and second as the cause of a cascade effect. With the collapse of the trout population in the Great Lakes, other invasive species that the trout fed on and effectively controlled, such as the alewife (a type of herring), increased rapidly. These exotics compete for the food of many native species and prey on their eggs and young. In turn, their voracious consumption of zooplankton, which grazes on phytoplankton (algae), caused heavy blooms of the latter, decreasing the clarity and health of the water. This deterioration of water quality in turn threatened a raft of native species.

In the realm of inadvertently introduced exotic species, the brown tree snake (*Boiga irregularis*) bears the greatest infamy. The snakes, native to northern Australia, New Guinea, Papua New Guinea, and the Solomon Islands, arrived on Guam shortly after the U.S. occupation of the island during World War II. Most likely the snakes were stowaways in military cargo (disabled materiel retrieved from the campaigns off Papua New

Guinea) that was unloaded at Guam's port after the end of hostilities. The only snake species native to Guam is very small, blind, and about the size of an earthworm; it feeds on ants and termites and is quite harmless. The brown tree snake, in contrast, is slightly venomous, arboreal, nocturnal, and a voracious predator. It can grow to ten feet in length. The local birds, bats, lizards, and small mammals of Guam had no experience with large, deadly climbing snakes or with any other predators that hunted nocturnally. As a result, the snake had a field day or, rather, a field night, in its new habitat.

The disappearance of bird species on the island followed in lockstep with the growth of the brown tree snake's geographical domain. By the mid-1960s, the reptile had colonized half of the island; the expansion was completed by the end of the decade. With an omnivorous diet, no natural predators, and adept skills, the brown tree snake thrived, reaching densities of 13,000 specimens per square mile—higher than snake densities in the Amazon and other native habitats. Once the island was fully in the snake's grasp, many species of birds, bats, and lizards fell victim. Indeed, the introduction of this serpent to Guam, ecologists assert, led to a series of extirpations that "may be unprecedented among historical extinction events in taxonomic scope and severity."[25]

The toll is high indeed. Of the eighteen species of birds native to Guam, seven have become extinct, two more are extinct in the wild, and nine other species have become rare or uncommon.[26] When birds became too sparse to sustain the appetite of the invader, it turned to reptiles. As a result, five species of native lizards have become extinct, in part owing to direct predation by the snakes and in part owing to the disappearance of birds that were crucial components of the lizards' diets. Many small mammals also suffered. Island shrews, for instance, are now rare.

Efforts to stop the carnage might be generously classified as too little, too late. Most native forest species of birds were virtually extinct by the 1970s, though they were officially listed as threatened or endangered by the U.S. Fish and Wildlife Service the following decade. In 1995, the USDA effectively upped the ante, listing the brown tree snake as one of the top three pests requiring control and eradication. This official reaction testifies to the ecological devastation. But the designation might also reflect the millions of dollars in damage caused to the island's electrical power lines by the serpent. Outages produced by the adept climbers occur every other day. Poultry farms are also regularly pillaged. To protect the domestic fauna and electrical equipment of Guam from the snake, thousands of barriers, traps, biocontrols, and toxicants have been put to use or are in development. Poisoned mice have been outfitted with parachutes and dropped into trees. Trained dogs are employed to ferret out snakes that seek passage from Guam to other islands as stowaways. Still, most of the damage done by the snake cannot be undone. The ecosystems of Guam have been radically transformed and likely will never recuperate.

Unfortunately Guam is not alone in seeing its ecological web of life destroyed by exotic serpents. Similar fates have befallen other island ecosystems, often because of the release of pet snakes. And slithering reptiles are not the only culprits. The list of invasive exotic fauna and flora ravaging islands, as well as mainland and aquatic ecosystems across the globe, is long and growing.[27]

Despite the Best Intentions

The destruction caused by overharvesting and exotic species is typically the product of shortsightedness. People simply fail to understand the long-term consequences of their biological tinkering. Such myopic choices often occur despite the best intentions. Conservationists, for example, have crafted policies to restrict the size of fishing nets so that only larger fish are caught, and hunting regulations often limit prey to full-grown animals. In both cases, the intention is to ensure that the young of the protected species are left to grow up and reproduce. But the effect of the regulations is counterproductive. Rather than seeking out the weak and small, like other predators in nature, human hunters and fishers are forced to pursue the large and strong of a species. The result is that the specimens that have demonstrated their ecological fitness are destroyed, and the species as a whole consequently suffers from a deteriorating gene pool. Indeed, there is evidence that some animals have adapted to our practices. Harvested and hunted species may now be reproducing fewer offspring at a younger age and, quite possibly, passing on traits not for fitness but for early reproduction.[28]

Probably the most famous example of good intentions paving the way to ecological mayhem occurred in the 1950s when the World Health Organization (WHO) attempted to quell an outbreak of malaria in Borneo. To fight the deadly disease, the WHO opted to spray DDT to reduce the mosquito population. The pesticide successfully controlled the malaria-carrying mosquitoes, but it also exterminated native wasps that fed on caterpillars, including a species of caterpillar that was fond of eating thatch. With the accidental extermination of the parasitic wasps, the thatch-eating caterpillar populations exploded. This proved most unfortunate for the Borneans, whose homes had roofs made out of thatch. Infested with caterpillars, the roofs started to leak and collapse.

To compound the problem, geckoes (a type of lizard) fed off the insects poisoned by the DDT. The geckos, in turn, were eaten by the local cats. Owing to the effects of bioaccumulation and biomagnification, many of Borneo's cats died. With fewer felines around, the population of rats grew markedly. And with rats running rampant on the island, agricultural crops were ravaged and outbreaks of typhus and sylvatic plague (diseases carried by the rodents) grew more common. The global health organization at this point called on the British Royal Air Force (RAF). Based in Singapore, the RAF came to the rescue by mounting Operation Cat Drop in which scores of presumably quite bewildered cats were parachuted onto the island from RAF planes.[29]

In retrospect, the connections become clear, and tracing them out makes for fascinating, though tragic, stories. The problem is that unintended consequences are mostly unanticipated. No one predicted the disappearance of moose from Vermont in the nineteenth century owing to the overharvesting of beavers. The moose, however, were dependent for their summer foraging on the vegetation growing in beaver ponds. With no beavers, there were no beaver ponds, no summer moose food, and no moose. Similarly, no one predicted the decline of songbird populations in the United States as a result of the extermination of cougars and wolves. But these large carnivores hunted midsize predators such as raccoons, foxes, and possums. And the burgeoning populations of midsize predators, with no cougars or wolves to prey on them, consumed too many songbirds and their eggs. While we may assuredly predict that unintended consequences will follow whenever we tinker with nature's web of relationships, we cannot predict what these consequences will be, when or where they will occur, or who and what they will affect.

The moral of the story is straightforward. Nature is a complex web of relationships. Plucking a particular strand out of this web inevitably produces reverberations across its breadth. Unanticipated and often dev-astating consequences may result, and good intentions provide no immu-nity. We cannot know the effect of our tinkerings, but we do know that they will surprise, and often sadden, us.

Negative Synergisms

Cascade effects highlight the tightly knit connections in nature's web of life and the chain of events that occur whenever a link is severed. But the unintended consequences of our actions are not limited to cascade effects. They can take other forms in natural systems, such as negative synergisms, discontinuities, and positive feedback loops.

Negative synergisms occur when several unrelated factors produce a greater impact as a result of their interaction than might be expected from simply aggregating individual effects. In other words, a perfect storm results from the unforeseen interaction of multiple contributing causes. In the ecological realm, negative synergisms typically arise when distinct stressors to individual species or ecosystems combine forces to yield dis-proportionately destructive results.

The severe damage suffered by the flora and fauna of many of eastern Canada's lakes and streams in the 1980s and 1990s is a case in point. The ecological destruction was a product of several factors—periods of drought, higher ultraviolet radiation, and sulfur dioxide emissions—none of which, taken singly, would have produced the level of damage that occurred. The droughts led to lower water levels. Droughts also reduced the sediment washed into streams and lakes, as there was little rain. This had the effect of increasing water clarity. The clarity was further heightened owing to acid rain, which increased the pH of the water. The high acidity of the rain in the region was caused by sulfur dioxide in the atmosphere, mostly from the emissions of coal-fired power plants. The very clear waters of streams and lakes were then subject to increased ultraviolet radiation from the sun, a product of lower levels of ozone in the stratosphere. The depletion of ozone was caused by the use of certain chemicals in previous decades, primarily chlorofluorocarbons (CFCs), which destroy the radiation-screening ozone molecules at high altitudes.

In the clearer streams and lakes, the high-intensity ultraviolet rays could penetrate up to 3 meters. Given the low water tables caused by the drought,

3 meters of ultraviolet penetration was often enough for the damaging radiation to reach bottom, where fish and amphibians lay their eggs. As a result, fish and amphibian populations, as well as other forms of aquatic life, suffered significant damage. A perfect storm of destruction had been produced by the interaction of three unrelated factors: drought, acid rain, and ozone depletion. Negative synergisms are not uncommon events. But they are typically discovered, if they are ever discovered, once the damage is done. Ecologist Norman Myers observes that negative synergisms may rule the day in our ecological future. "Probably the biggest environmental problem of all on the horizon," Myers writes, "will turn out to be that of the interactions between lesser problems."[30]

Discontinuities

Like other ecosystems, lakes have adaptive properties that mitigate change. For instance, the acidification of lakes caused by sulfur dioxide emissions may be slowed or wholly offset owing to the buffering capacities of water systems. At a certain point, however, the buffering ceases. Acidification then spikes, and a lake can quickly die. Whereas the sources of acidity may have remained relatively constant, the acidification that once proceeded slowly and in a predictable way suddenly rises, transforming the entire ecosystem. Such an unintended consequence is a discontinuity. It occurs when a typically unperceived threshold gets crossed, after which ecosystem dynamics no longer operate in a linear fashion. At that point, causes cease to produce proportional effects. Discontinuities mark an abrupt shift in a trend, the abruptness and severity of which could not have been predicted from previous patterns. After the threshold is breached, the change that the ecosystem undergoes often proves irreversible. The damage done may be too intensive or too pervasive to repair. The collapse of Canada's cod fishing industry in 1992 is a prime example.

At the time of the European discovery of North America, seafarers plying the Grand Banks—the 300-mile continental shelf off the coast of Newfoundland—told stories of cod so plentiful they could scoop up the fish in wicker baskets. Explorer John Cabot marveled in 1497, probably with some hyperbole, that teeming schools of cod virtually blocked his ships. Salted cod quickly became a staple in Spain and Portugal, whose fleets returned again and again to exploit the aquatic riches of the New World.

Cod fishing became a key industry of the region. At first, primitive but effective traps were employed to corral and catch the olive green and spotted omnivorous fish, which average from five to ten pounds each and occasionally weigh twenty times as much. Longlines were also used, as were gill nets. The big change came in the early 1950s as massive factory trawlers from nations across the globe arrived in the cold waters off Canada's coast, dragging nets the size of football fields behind them. These trawlers plied the seas to within 12 miles of the shore, catching, processing, and freezing fish around the clock for weeks or months at a time. Annual catches of cod from the region rose to 800,000 tons.

When annual catches declined to 300,000 tons in the mid-1970s, Canadian officials extended their territorial waters to 200 nautical miles offshore, effectively banning foreign trawlers from the Grand Banks. Canada's fishermen worked feverishly to pick up the slack, outfitting their own factory trawlers and setting to work. But something was awry. By the late 1980s, Canadian fishermen were finding it difficult to catch as many fish as they were allowed, even though government officials had reduced quotas in response to diminishing yields. Then, in 1992, it all came crashing down: the fish were nowhere to be found. Canada's minister of fisheries and oceans was forced to issue a total ban on the fishing of northern cod. Over 40,000 fishermen and employees of the processing industry were thrown out of work and entire communities collapsed. Notwithstanding the ban, the northern cod population continued to decline, with the total biomass falling to 1,700 tons in 1994, a decline of almost 400,000 tons over four years. There would be no recovery. The North Atlantic cod was officially listed as an endangered species in 2003 and remains so today.

The cause of the collapse is still debated. Normally one might expect fishing stocks to decline in a linear fashion, in proportion to increases in annual catches. Efforts to manage the cod industry by determining the maximum sustainable yield, it was believed, would maintain an equilibrium. Lower the quota of fish that can be caught and, with more fish left in the sea, stocks would surely rebound. Quotas could then be slowly raised to take advantage of the increased numbers of fish. That is the logic, but it is not what happened.

Most likely the methods employed in catching the fish were at fault. The enormous factory trawlers sported huge dragger nets equipped with rock hopper dredges. These nets, some with 3,500-feet circumferences, were held open by steel doors, heavy chains, and rollers that bumped and scraped along the ocean floor, netting everything in their path and causing extensive damage to the sea-bottom community. Juvenile cod thrive in rough terrain, hiding from predators amid the nooks and crannies of a variable seafloor. The highly efficient trawling equipment destroyed the habitat from which depleted stocks might otherwise have rebounded.

Interactions among species may also have played a role in the collapse. Extensive mackerel fishing probably caused a rise in herring populations, as herring are one of the chief prey of mackerel. Since herring feed off cod eggs, fewer mackerel meant more herring, which meant fewer cod eggs. Whatever the precise confluence of contributing factors, by 1992 a threshold had been crossed and a tipping point reached. Then the cod were gone. In the realm of ecological communities, the problem with tipping points is that we seldom see them coming.

Positive Feedback Loops

Positive feedback loops, also called vicious cycles, occur when a noxious trend effectively reinforces itself, increasing in severity at an accelerating rate. The phenomenon of climate change provides all too many examples of positive feedback loops. The planet is steadily warming owing primarily to increased levels of anthropogenic (human-caused) heat-trapping gases in the upper atmosphere. Carbon dioxide, the major greenhouse gas, has climbed from less than 280 parts per million (ppm) in preindustrial times to 390 ppm in 2011. It is steadily rising. This has caused an increase in average global temperature of over half a degree Celsius. Snow and ice do not take kindly to warmer weather, so the arctic and Antarctic ice fields have been shrinking at an alarming rate. Indeed, predictions are that the arctic may be ice free in summer as soon as 2030.[31] That is good news for shipping companies, some of which are already preparing for the opening of a northern passage. But it is bad news for polar bears and other members of the arctic ecosystem that depend on sea ice as a crucial feature of their habitat. It is also bad news for nations across the globe whose people depend upon river and lake waters fed by now quickly melting glaciers. And it is very bad news for island and coastal dwellers because all the melted snow and ice, primarily from Greenland and the continent of

Antarctica, will cause the oceans to rise by as much as 1 to 2 meters by the end of this century.[32]

Perhaps most disconcerting, the melting of polar and glacial snow and ice is also bad news for the stability of the earth's climate. That is because the shrinking of ice fields, itself a product of global warming, will accelerate the rate at which further planetary warming occurs. This warmer climate means that even more ice will melt, producing a vicious, self-reinforcing cycle of shrinking ice and rising temperatures. The science is elementary. Ice and snow provide reflective surfaces, casting back into space much of the incident light that strikes. They have what is known as a high albedo. Open water, and even more so exposed earth, has a relatively low albedo, as do all darker surfaces. Melting snow and ice translates into less sunlight reflected back into space. That means more sunlight, and more of its accompanying heat, will become absorbed by the oceans and the earth's surface. To make matters worse, warmer waters release more of their dissolved carbon dioxide into the air than do colder waters. Ice-free waters mean that more greenhouse gases will be released into the atmosphere from the oceans themselves. The upshot is alarming: the faster the planet's ice melts owing to global warming the faster will the planet warm.

Climate change produces a similar vicious cycle in another part of the cryosphere, that portion of the earth's surface where water is in solid form. The subsurface of high northern and southern regions of the globe, up to 20 percent of the planet's landmass, is frozen year round. Although a thin, active layer of soil lying at the surface may thaw in summer and come alive with a profusion of plant and animal life, the deeper soils, up to 4,500 feet deep in some regions of Siberia, remain permanently frozen. With warmer ambient temperatures, increasing amounts of this permafrost will thaw in discontinuous regions, that is, areas where ground temperatures normally range within a few degrees of the melting point. That will be devastating for ecological communities that have adapted to a permafrost habitat. And like the vicious effects of shrinking ice fields, the melting of permafrost will produce its own positive feedback loop. In this case, however, albedo has nothing to do with it. The problem is that frozen soil contains a great deal of trapped carbon dioxide and methane, the latter being an even more potent greenhouse gas. As long as the ground stays frozen, these trapped gases stay put, where they can do no harm. As soon as the permafrost thaws, the gases are released into the atmosphere.

Thawing tundra in Siberia alone could pump many trillions of pounds of carbon dioxide and methane into the atmosphere, heightening the greenhouse effect and greatly accelerating global warming.

Unfortunately, still more vicious cycles plague the phenomenon of climate change. First, consider the positive feedback loops between vegetative cover and hydrology. The Amazon basin contains up to 30 percent of the biological diversity of the planet. It boasts 20 percent of the earth's surface freshwater and its largest contiguous tropical forest. Most of the water in the Amazon region is recycled, as up to 75 percent of the rainfall returns to the atmosphere in the form of evapotranspiration from vegetation. In the near future, there will likely be a significant climate-driven dieback of the Amazonian rain forest as a result of regional drying. Global warming can produce regional droughts, and drought causes tropical forests to recede. In turn—and this is where another vicious cycle arises—receding forests have the effect of exacerbating drought conditions. This occurs for a number of reasons.

First, there is a biogeophysical feedback loop wherein reduced forest cover suppresses local evaporation, and less evaporation means that water does not get recycled into the atmosphere, so there is less rain. Many forests have trees with roots that tap the groundwater. If these forests have dark canopies (low albedo), they will also absorb heat and hence increase evaporation rates. This will produce rising water vapor, which releases heat as it condensates in cumulus clouds, causing showers and effectively creating a rainy microclimate. In turn, forested lands covered with perennial plants and their deposited litter allow rain to be absorbed by the topsoil, where it may be taken up directly by vegetation or seeps into groundwater. In contrast, deforested lands, particularly in hilly or mountainous regions, lose much of the rainfall they receive to runoff. This runoff ends up in rivers, lakes, and the ocean, with less of it available to reinforce the rainy microclimate of the region. Moreover, heightened ambient carbon dioxide concentrations cause plant stomata (the pores on the plant epidermis that control the passage of water vapor) to open less. This reaction to higher carbon dioxide levels reduces the release of moisture from forested areas to the atmosphere, further decreasing water cycling and further drying the microclimate.

And things get even more vicious. A biogeochemical feedback loop occurs when the formerly forested lands recede. As the trees and biomass of the Amazon die, they either slowly decompose or, once they dry out, quickly burn in wildfires. Both their decomposition and their burning release carbon dioxide into the atmosphere, effectively accelerating global warming trends. Smoke from fires (including those deliberately set to clear land) also contributes to a decrease in local rainfall. Finally, warmer soils accelerate decomposition, and the decomposition of biomass releases carbon dioxide and methane. Positive feedback loops, then, are initiated by both climate-induced drying and the deforestation that is caused by agriculture, timber harvesting, and human settlement. A negative synergy then develops between these two phenomena.

In sum, global warming will likely decrease tropical forest cover, promote regional drying, decrease carbon intake from vegetation (after a threshold is passed), and increase carbon release from decomposing and burning biomass and denuded soils. All this will exacerbate global warming, which will make the cycle still more vicious.

Vegetated wet states and barren dry states are relatively stable and self-reinforcing ecosystems. Each perpetuates the wet or dry microclimate that sustains their respective level of vegetative cover. As the macroclimate changes, formerly wet forested areas may reach a hydrological threshold, followed by an abrupt shift to another equilibrium (a nonlinearity). The result will be the creation and expansion of barren, dry lands that will not easily, if ever, become reforested. If current trends continue, a tipping point might be reached by the middle of this century that turns the Amazonian rain forest, formerly one of the planet's largest carbon sinks, into a net carbon releaser. Similar vicious cycles are being generated on the forested lands of other continents.

Scientific studies that do not account for negative synergies, positive-feedback loops, and other nonlinearities may be underestimating the extent, level, and speed of global warming by as much as 50 percent. With this in mind, many climatologists reject models grounded on "straightforward cause–effect relationship[s]" and encourage their colleagues to integrate "nonlinearities at disparate scales."[33] The complex interdependencies between atmospheric carbon levels, ambient temperatures, hydrology, carbon uptake and release, vegetative cover, and microclimates is enough to dizzy the most sophisticated scientist. However, to get a reasonably accurate picture of our dynamic world—and work effectively to preserve it—climatologists realize that we can never study just one thing.

Systems Thinking

Ecologists focus on interdependencies within biological habitats and the consequences when a particular linkage is altered or severed. These consequences may be direct and obvious, or indirect, difficult to observe or anticipate, and mediated by extensive chain reactions, feedback loops, negative synergies, or discontinuities. Conservation efforts grounded in ecological science therefore must not only measure and mitigate the direct effects of human actions on threatened species; they must also grapple with indirect effects. The U.S. Endangered Species Act (1973) was originally designed for just such purposes. It not only proscribes or limits "takings" of endangered species, that is to say, their hunting or harvesting; it also proscribes or limits actions that cause a decline in an endangered or threatened species, even if that action does so unintentionally or indirectly. The act of draining a wetland or clear-cutting a forest, for instance, may not in itself kill migrating birds that sustain themselves seasonally in these habitats. But it will surely contribute to their decline, assuming suitable alternate habitat is in short supply.

In 2008, during the waning months of the presidency of George W. Bush, the definition of "indirect effect" in the Endangered Species Act was slated for revision. It would now designate only those effects on wildlife populations for which the proposed human action constituted an "essential cause," defined as an action that is "necessary for the effect to occur. That is, the effect would not occur 'but for' the action under consultation and the action is indispensable to the effect."[34] This change in language would have made it much more difficult to proscribe or limit actions that contribute to the decline of a species.

Endangered species typically face multiple threats, such as habitat loss, competition from invasive species, pollution, overharvesting, and climate change. No single contributing factor is necessary for a decline in a species to occur, as it might well take place (at a diminished rate) even if one of the contributing factors were not in play. Stripped of its ability to proscribe or limit human actions that are not solely responsible for the decline of a species, the newly worded Endangered Species Act would be hamstrung. Clearly the proposed change in language flew in the face of ecological reality, which seldom, if ever, exhibits monocausal relationships and always remains susceptible to negative synergies.

In one respect, however—presumably not that intended by the Bush administration—doing away with the language of "indirect effects" makes ecological sense. To employ the terminology of direct and indirect effects, or to distinguish between effects and side effects, is unecological. Garrett Hardin rightly observes that

when we think in terms of systems, we see that a fundamental misconception is embedded in the popular term "side effects." . . . This phrase means roughly "effects which I hadn't foreseen, or don't want to think about." As concerns the basic mechanism, side effects no more deserve the adjective "'side" than does the "principal" effect. It is hard to think in terms of systems, and we eagerly warp our language to protect ourselves from the necessity of doing so. . . . The dream of the philosopher's stone is old and well known and has its counterpart in the ideas of skeleton keys and panaceas. Each of these images is of a single thing which solves all problems within a certain class. The dream of such cure-alls is largely a thing of the past. We now look askance at anyone who sets out to find the philosopher's stone. The mythology of our time is built more around the reciprocal dream–the dream of a highly specific agent *which will do only one thing*. . . . [But] wishing won't make it so. . . . *We can never do merely one thing*.[35]

Hardin's point is that the world of side effects is a human conceit. It is a place where our intentions and motivations are privileged. Collateral damage that occurs as an unintended consequence of action becomes discounted or dismissed.

In nature, a complex web of interdependence that operates without intentions or purposes, there are no side effects. There are causes and there are effects. And every effect is also a cause (of subsequent effects). And every cause is also an effect (of earlier causes). The earth is free of unintended consequences. Only the world—the human *oikos*—is rife with them.

Consider the *oikos* of gray squirrels and oak trees. When acorns are ripe for collecting in the fall, the squirrels find themselves in quite a dilemma. They cannot possibly consume all the available nuts, and the approaching winter will bring lean times, so the squirrels have invented a familiar practice: storing harvested acorns underground and in the nooks and crannies of trees. In winter, these caches provide much needed sustenance. A single squirrel may collect and store many hundreds of nuts in a season. But how do these resourceful rodents retrieve so many nuts, many, if not most, of which are hidden away as singletons in a process called scatter-hoarding? Squirrels may employ their sense of smell to help them find their caches and have been seen rubbing acorns to apply scent before stashing them. For the most part, however, they employ an extraordinarily acute spatial memory.

The squirrel's memory is amazing. Imagine trying to remember to within a few centimeters where you buried many hundreds of different items in a forest months earlier. But squirrels are not miracle workers. Their spatial memory is good but far from perfect. Studies have shown that squirrels may remember only where a quarter of their nuts are stashed. That is why they need to bury hundreds of them each season. In the end, they will retrieve and eat only a portion of their nuts.

Now this all seems highly inefficient, as squirrels must spend tremendous amounts of time and energy collecting and hiding hundreds of nuts each year, only to leave many, if not most, of them in the ground. But squirrels, no less than humans, can never do just one thing. So collecting and hiding acorns is not simply making provisions for hungry winter nights. It is an act of forestry. Every acorn that does not get retrieved and eaten stands a decent chance of becoming a sapling. Saplings grow into trees, and an oak tree provides a home with a built-in restaurant. The forgetful rodents are sustaining their *oikos*.

Oaks take a long time to mature. By comparison, squirrels do not live very long (typically less than six years). So the squirrel that buries but does not retrieve scores of acorns each year is not planting oaks for itself. The benefits of its inefficient industry will neither satisfy its hunger nor provide it with an arboreal abode. Effectively, it is planting oaks for the benefit of its great-great-great-grand-squirrels. Of course, that is not the squirrel's motivation, intent, or purpose. The rodent is simply storing nuts for a future meal. Were squirrels better at finding their caches, however, there would be fewer oak trees, and hence fewer squirrels, within a relatively short span of time.

From an ecological point of view, it is illegitimate to deem the abandoning of acorns in the ground a side effect of the squirrels' harvesting and hording. This act sustains the habitat and food source that squirrels rely on. It is as important to the species as the consumption of nuts. And from the oak tree's point of view, nourishing squirrels might best be viewed as the side effect of its own efforts to self-propagate. Effectively, oak trees use forgetful squirrels as acorn planters.

From the point of view of the forest ecosystem, the distinction between effect and side effect makes no sense whatsoever. The ecosystem is reproducing itself through the complex interactions of squirrels and oaks. From a systems point of view, every cause has multiple effects, none more or less privileged than any other, and every effect is itself a cause, another stone sending ripples across the pond of life.

Ecosophic awareness entails thinking both in terms of the many interdependent parts of a system and, simultaneously, in terms of the whole. The first and easiest task is to think of the parts. Within biological systems, this entails investigating the role of particular species and their interactive relationships. The second step, equally necessary but more difficult, is to consider the whole. This is what Leopold was suggesting when he asked us to think like a mountain. To think of the whole is to engage in systems thinking. From the point of view of the whole, there are no side effects because there are no sides, no peripheries, no out-of-bounds. There is the whole, the eco*system*. In our example, there is a web of relations between squirrels, oaks, soils, insects, fungus, other parasites and predators, forest fires, rainfall, air, sunlight, and much more. There are multiple chains of causes and effects, and multiple relationships between different species that allow any individual species to survive and prosper. From a systems point of view, the web of relationships rather than individual entities claims our attention.

The very notions of indirect effects, side effects, and unintended consequences are products of our taking a nonsystemic perspective. Actions can have side effects only if we assume that their consequences are unintended (in the case of humans) or are relatively less important to immediate goals or pursuits (in the case of other species that do not exhibit intentional states of mind). But as deep ecologist Arne Naess maintained, ecological reality is not composed of "completely separable objects" that harbor intentions, purposes, or goals. Rather, it is composed of fields of interaction wherein all things "hang together."[36] From an ecological perspective, Naess observes, individuals are not separable objects but rather "knots in the biospherical net or field of intrinsic relations."[37] Relationships—not individuals, their intentions, or their particular actions and effects—are foremost. Relationships extend in all directions.

An ecosystem might be likened to a piece of choral music. Harmony arises when all the voices are heard together. It is not to be found in the individual notes or voices, as beautiful as they may be, but in the resonance that occurs among the notes and voices. To think like a mountain is to perceive this resonance. It requires us not only to see the whole as composed of interactive parts but to understand the individual parts as reverberating to the beat of the whole.

To speak of "the whole" is always misleading, for any whole that we aspire to capture in our thoughts or words can always be located within another more encompassing framework. The nineteenth-century psychologist and philosopher William James observed that regardless of how "vast or inclusive" it might be, the whole we attempt to grasp or depict can never be *the* whole. "The word 'and' trails along after every sentence. Something always escapes," James insists. "'Ever not quite' has to be said of the best attempts made anywhere in the universe at attaining all-inclusiveness."[38] Systems thinkers are well aware of this caveat. They acknowledge that to think from a systems point of view only ever provides "a partial description of a reality based on one observer's perspective." At best, one's "limited mental representation" manages to gesture at the whole and its multifaceted complexities.[39] To think like a mountain, from a systems point of view, is an aspiration not an achievement.

Perhaps Leopold's recommendation already acknowledges this inevitable limitation. To think like a mountain is to escape the restricted purview of an isolated species to better reflect the mountain's point of view. But a mountain sits within a mountain range. And a mountain range exists within a still more expansive continental geography. And continents share the globe. Thinking like a mountain is a supreme challenge. But something always escapes.

A Coevolutionary Perspective

Just as one's geographical whole can always be placed within a more encompassing spatial realm, so one's time can always be placed within a broader temporal realm. Thinking like a mountain means thinking through the countless ages that a mountain has witnessed as it rose from tectonic forces to claim the sky. To think like a mountain is not to produce a snapshot of a mountain ecosystem, however panoramic one's purview. It must also be temporally expansive. Systems thinking grapples with a web of relationships in the here and now *and* in its transformations over time. A coevolutionary framework is employed. The assumption is that the evolutionary paths of species maintaining close ecological relationships with each other largely depend on the shifting patterns of their interactions.

In 1859, Charles Darwin first addressed the topic of "coadaptations."[40] A century later, Paul Ehrlich and Peter Raven coined the term *coevolution* to describe the mutual adaptations of interactive species.[41] Evolutionary change is not simply a matter of isolated species adapting themselves to their respective ecosystems. Rather, it is the product of multiple species mutually influenced through extended interactions. A coevolutionary framework captures the adaptive changes in relationships that develop among interdependent organisms. These organisms do not simply occupy habitats: they transform them. In turn, they are themselves modified by the habitats that they help to constitute and transform. Coevolution is grounded in the "interpenetration" of organism and environment.[42]

Coevolutionary frameworks incorporate the temporal into systems thinking. The parts—individual species—are understood as evolutionary products of the whole. The whole in this instance is the ecosystem over time, a dynamic environment defined and modified by relationships of interdependence among the species that make it home.

Musical improvisation rather than choral harmony is the better metaphor to capture this phenomenon. In an ecosystem, no conductor is orchestrating individual efforts into a unified whole toward a predetermined finale. The harmony is dynamic, not static. It is open-ended, not fixed. And it is not a product of top-down planning. Rather, it is the result of interactive efforts that are responsive to systemwide effects. The individual components of an ecosystem participate in a constant jam session. They are like jazz musicians who make music by individually responding to a collectively created resonance. To be sure, there are underlying patterns: particular chord progressions, preexisting songs serving as themes, and adopted keys that set the stage for interaction. Likewise, patterns of predation, competition, and mutualism between species in the context of the struggle for adaptive fitness provide the foundations for ecosystem dynamics. But dynamic, open-ended, coadaptive interaction is the name of the game. As Bob Marley observed in his song *Jamming*, "Ain't no rules, ain't no vow, we can do it anyhow."

The whole exists only owing to the contribution of each of the parts. Yet each of the parts does what it does in response to the whole. The parts are nodes on Indra's net, defined by and reflective of each other. Strands and jewels in this web of life are regularly added and rearranged. A jam session never completes itself as a written piece of choral music is completed once the last note is played or sung. Jam sessions end, to be sure. But their ending marks the termination of something that might well have continued, and continued to transform itself. New chord progressions might have been added; other players and instruments might have joined in. Ecosystems, like jam sessions, can never be captured conceptually without artificially truncating their spatial extension and temporal dynamism. Something always escapes.

In his book *The Web of Life*, Fritjof Capra writes, "Understanding ecological interdependence means understanding relationships. It requires shifts of perception that are characteristic of systems thinking—from the parts to the whole, from objects to relationships, from contents to patterns."[43] Capra rightly characterizes the dynamic nature of such relationships in terms of coevolutionary processes. Ecosystem interdependence is a freeze-frame snapshot of inherently dynamic events. Coevolution is the temporal meaning of ecological interdependence. It is what ecological interdependence looks like in time.

There Is No *Away*

Michael Soule, one of the foremost scholars of conservation biology, observed that "every personal act—of production, consumption, travel, communication, recreation, disposal, voting—is an ecological act. And every ecological act should be a conscious one."[44] The thought is profound and daunting. Since everything is hitched, each of our actions must be engaged with an eye to its ecological effects and, to employ the human conceit, with an eye to its multiple side effects. Discerning and predicting these effects and side effects is anything but elementary. Wendell Berry, a farmer, writer, and poet, observes that ecologically, "there may be specialized causes, there are no specialized effects." The effects of our actions, Berry continues, are "invariably multiple, self-multiplying, long lasting, and unforeseeable in something like geometric proportion to the size or power of the cause."[45] When we act into nature, we inevitably begin or transform processes we do not foresee and cannot control.

To say that everything is connected is not to say that everything is connected to everything else in the same way. Some relationships are strong, direct, and enduring. Others are weak, mediated, and fleeting. We will not always know which is which. Ecologists, it follows, "should expect unsuspected and surprising connections to be important with some regularity."[46] The fact that we live not only within webs of ecological relationships but also within webs of social relationships makes things even more complex. Personal acts are, almost inevitably, both social and ecological acts. Within this context, the regularity of surprising connections should not surprise us at all.

In 2001, when storms in Inner Mongolia carried 50,000 metric tons of dust across the sea, dropping it in the United States, few would have thought to blame their own penchant for soft sweaters. Yet the foreign dust settling on Americans—more than twice as much dust as is produced domestically in a typically day—was whipped up in storms raging over eroded Asian grasslands. The erosion was largely caused by the vast increase of goats whose fur was used to make cashmere sweaters, mostly sold in the United States.[47] Buying a cashmere sweater—even with the good intention of wearing it in lieu of turning up the thermostat so as to lower one's carbon footprint—is not doing just one thing.

Unintended consequences may follow as regularly from the best-intentioned actions as from the most self-serving. Consider the act of putting trash where it belongs—in the trash bin. When the contemporary environmental movement was inaugurated in the 1970s, campaigns to put an end to then widespread littering were prominent. The most common environmental message was, "Don't be a litterbug." By any measure, the antilittering campaign was a stunning success. Within a generation, littering was transformed from a common practice to a despicable crime. At the time, however, few people asked, "And then what?" once they dutifully threw their trash in the bin. Keep America Beautiful, a coalition of companies involved in glass, aluminum, paper, and plastic production and use, ran a popular TV spot in 1971 depicting roadside litter and a Native American actor, Iron Eyes Cody, with a tear running down his cheek. It was highly effective. Meanwhile, Keep America Beautiful was actively engaged in opposing a national bottle bill that would have mandated a recycling deposit for glass bottles. Few made the connection or saw the contradiction.

Still, there were visionaries who knew better. In 1971, biologist Barry Commoner wrote *The Closing Circle*. *Time* magazine dubbed Commoner "the Paul Revere of Ecology." He is often identified with Rachel Carson as the coparent of the contemporary environmental movement. *The Closing Circle* identified the central "laws of ecology." The first law, according to Commoner, is that "Everything is Connected to Everything Else."[48] The second law, which follows from the first, is that "everything must go somewhere. . . . Nothing 'goes away'; it is simply transferred from place to place, converted from one molecular form to another."[49] A Greenpeace television advertisement years later illustrated Commoner's two ecological laws in a powerful fashion that brought the antilittering campaign to a new level. The advertisement presented the image of an overflowing landfill. When we next throw away a piece of trash, Greenpeace enjoined viewers, we should ask ourselves, "Where is 'away'?"

Tidy streets lined with trash cans may mark the end of one problem. And litter surely is a problem worth solving. But brimming trash cans constitute the beginning of another problem. Garbage trucks use lots of fossil fuel. Landfills and trash incinerators are not environmentally benign, often leaching toxins into watersheds and polluting the air with emissions. Garbage chucked in the bin may be out of sight and out of mind, but it is not, ecologically speaking, gone. There is no away.

The "And then what?" question that ecologists prompt us to ask can usefully be posed in reverse. Regarding our example, we need to question how there came to be so much litter to dispose of in the first place. When next opening the lid to the trash can to get rid of the plastic cradling our latest purchase, we might interrogate the commercial culture that stimulates the overproduction of disposable goods, excessive packaging, and hyperconsumption. From here, we might observe the high carbon footprint that is produced by our reliance on fossil fuels in the manufacture and global transport of products and packaging. So the next time we throw out a piece of trash, there is a lot to think about: that our purchase may stimulate manufacturers to produce more disposable, overpackaged, or nonrecyclable goods and that our discarded trash now begins a journey with significant environmental costs.

Littering should not be tolerated, but it visually brings home the fact that there is no away, and that a society addicted to consumption and waste is an ugly place. If one day each year everyone dumped their trash on Main Street for all to observe, wade through, and smell, we would have a valuable lesson in systems thinking. After all, every personal act—from what we make, to what we buy, to what we throw away—is an ecological act.

Conclusion

In a world that demonstrates "the fundamental interdependence of all phenomena," Fritjof Capra observes, the most needful thing is an "ecological perspective."[50] The fabric of life is woven of relationships. It harbors networks of such complexity that no amount of investigation can fully reveal the intricacies of the patterns or the eventual consequences of our plucking at strands. We should always ask, "And then what?" before we act. But at the end of the day, we must realize that our persistent questioning will never exhaust the relationships we are making and breaking, starting, stretching, and severing. Ecologically speaking, Wendell Berry observed, we have "never known what we are doing because we have never known what we have been undoing."[51] That is a truth that will become more patent, not less true, with each advance in our knowledge.

Nature is deeply complex, more complex than we can hope to know. We can aspire to think like a mountain, to think holistically from a systems point of view. But as individual parts, we can never truly know the whole. What we do know is that the world is transformed with each of our actions and that we, in turn, are transformed by it.

Ecosophic awareness entails a humble acceptance of the limitations of our knowledge *and* a hopeful embrace of the awesome responsibility to safeguard the fabric of life. Though daunting, the prospect of participating in a whole that we can never fully know or control need not debilitate. Quite the opposite. The more we understand the web of life, and our place within it, the more we may feel ourselves at home. Our earthly home is an *oikos* where we engage in vibrant relationships and become transformed by our engagement. Existing within this interdependent world, the best option available to us, as for any other species, is to coevolve.

One might summarize the examples and arguments presented in this chapter with a simple phrase: nature bats last. This is not to suggest that we, as puny humans, are condemned to failure and futility. To be sure, we are not sovereign rulers exercising absolute control. Our greatest achievements will always produce unintended consequences. That is what it means for nature to bat last. The important thing to recognize, however, is that nature built the stadium and produced all the athletes. Nature does indeed bat last. But at the end of the day, we are on the same team. We should play as such.

2 Ethics

Out beyond the idea of right doing and wrong doing, there is a field. I will meet you there.

—Rumi

Ethics concerns relationships of reciprocation, obligation, and caring within communities. It is generated in response to the affective bonds we have to others and our reasoned understanding of the requirements of social order, justice, and wellbeing. Humans are not the only creatures on this earth that develop norms and values; communal mammals such as great apes, for instance, appear to value, promote, and enforce reciprocity. However, our species has most fully developed its ethical habitat—to the degree that it affects most, if not all, of our cultural practices.

Ethics is a field of study and a facet of life occupied with our responsibility to sustain the community that sustains us. Today, however, the community that sustains us is no longer the extended family, tribe, clan, village, or even nation state. What does ethics mean, then, given our planetary interdependence? What ethical obligations do we have in a global age?

My own experience of the global reach of ethical relationships came just after my junior year of college when I was in an exchange student in Haiti. As an introduction to the frequently exploitive economic relations between developed and developing countries, we were shown a documentary film, *Guess Who's Coming to Breakfast*.[1] The film outlined how the cereal we eat each morning with bananas effectively invites the multinational corporations that produced these foods to our breakfast table. These corporations, which we effectively endorse when we buy their products, may carry out socially and ecologically devastating practices in developing nations while supporting repressive, undemocratic political regimes. Eating cereal each morning, the documentary suggested, implicates us in this economic exploitation, ecological destruction, and political oppression. Having breakfast is not doing just one thing. Our morning nourishment and gustatory pleasure come at a high price to others. The fact that these victims are unseen and unheard makes our dining bearable but no less unethical.

Until that time, I had believed my ethical actions and responsibilities began and ended with the people I influenced immediately and concretely by my words and deeds. My experience in Haiti shattered that sense of isolation. I began to ask, "And then what?" whenever I shopped, ate, or went about my daily routines. The world came alive with a dizzying array of ethical relationships mediated by technology, politics, and economics. The following year, back in college, I joined a student-led antiapartheid group whose purpose was to convince the university administration to divest all of its financial holdings in corporations that carried out business with the then racist regime of South Africa. We held rallies, shouted slogans, and carried placards denouncing the university's "blood money." We demanded the immediate release of Nelson Mandela, the antiapartheid activist and leader of the outlawed African National Congress who was imprisoned on Robben Island.

That was in 1980. A decade later, Nelson Mandela, after twenty-seven years in captivity, was freed. In another four years, the first multiracial elections were held in South Africa, Mandela became president, and racial segregation was outlawed. I would like to think that my small agitations as a college student contributed to the end of the brutal regime of apartheid in South Africa. But I could not begin to demonstrate any direct cause-and-effect relationship. And if my "contribution" to the end of racist laws and policies made me feel that I was on the right side of history, a growing awareness of my less beneficial, albeit indirect, involvement in the fates of people across the globe gave me pause. When I dug beneath the surface of my daily transactions, I found countless culpabilities. Indeed, in today's global economy, most purchases of goods and services, as well as many other actions and nonactions, link us economically, socially, ecologically, and politically to systems of repression, exploitation, and degradation around the world.

Consider how the morality of our breakfasts (and lunches and dinners) are complicated by climate change. Currently almost one-third of all greenhouse gas emissions come from land use, which is to say from deforestation and agriculture. A major culprit is the use of artificial fertilizers (made from petroleum), which produce copious amounts of nitrous oxide, a greenhouse gas three hundred times more potent than carbon dioxide. Organic farming yields significant reductions in greenhouse gas emissions. So buying locally grown organic produce may significantly reduce our personal contribution to climate change. In turn, livestock accounts for half of all the greenhouse gas emissions produced from agriculture and land use. The suffering of livestock raised in factory farms is an ethical issue for many people. But even beyond this important concern, refraining from or reducing the consumption of meat might be considered an ethical imperative. The fact of the matter is that vegetarian diets significantly reduce greenhouse gas emissions.[2] Indeed, what we serve for breakfast, lunch, and dinner has as much impact on climate change as what vehicles we drive or what energy sources we use to heat, cool, and light our homes. If mitigating climate change is an ethical obligation, then the foods we eat and the means we employ to produce them are moral matters.

Virtually every action we take, or fail to take, has ethical and ecological implications. But if we posed and attempted to answer the "And then what?" question before each of our actions and transactions, we would be left with no time to do anything else. The effort to trace the expanding chains of causes and effects, chart the accompanying network of relationships, and discern the justice or injustice of interactions would consume our lives. And at the end of the day, we would be juggling probabilities and skirting uncertainties. The mandate to determine the full ramifications of action and ensure its moral purity would lead to paralysis.

Passivity and paralysis in the face of our ignorance is both dangerous and ethically suspect. In an increasingly globalized world of accelerating technological development, social injustice, political repression, and ecological destruction, sins of omission may prove as injurious as sins of commission. Benjamin Franklin said that "the man who does things makes many mistakes, but he never makes the biggest mistake of all—doing nothing."[3] Franklin had a point but misspoke. Doing nothing is never really an option. We cannot escape ethical obligation without quitting life itself. Nonaction has effects and side effects, many of which are ethically significant. At the Holocaust Museum in Washington, D.C., an inscription pungently drives home this fact. In tribute to the 6 million Jewish victims of the Nazi regime, the inscription reads: "Thou shalt not be a victim. Thou shalt not be a perpetrator. Above all, thou shalt not be a bystander."

Bystanding has consequences. To forgo the pursuit of justice is often to participate, by way of passivity, in regimes of injustice. It is said that "the only thing necessary for evil to triumph is for good men to do nothing." Martin Luther King Jr. voiced this conviction with eloquence. In his 1963 "Letter from a Birmingham Jail," King wrote, "Injustice anywhere is a threat to justice everywhere. We are caught in an inescapable network of mutuality, tied in a single garment of destiny. Whatever affects one directly, affects all indirectly."[4] On Indra's net—our single garment of destiny—injustice pertains not only to things done but also to things left undone. Our legacy to life is found in what we do and what we fail to do. Stopping to think about the effects of our own actions in a connected world is the first requirement of ethical life. It is a daunting task, but it does not release us from the responsibility to act.

This disturbing conundrum takes us to the vexing question of how an ethical theory of interdependence—a sustainability ethic—can effectively be put into practice. Sustainability serves self-interest. It concerns safeguarding goods for the long term—things we need and want and cherish. Sustainability is also an ethical conviction and practice. It goes beyond safeguarding goods and satisfying long-term self-interest to address the pursuit of the good life. The good life is an ethical ideal.

Ethics is treacherous ground. Many have sought to navigate its labyrinths, and many have become disoriented and lost. To guard against aimless wandering as we explore the moral terrain of a connected world requires observing some well-established guideposts. In Western thought, the ancient Greeks were the trailblazers of ethics. They provided the earliest and, many would argue, the most enduring beacons. From them we will learn that ethics, like sustainability, is a dance with uncertainty. We then turn to the eighteenth-century German philosopher Immanuel Kant, still one of the most cited figures in moral philosophy. Kant demanded greater resolution from ethical frameworks. To avoid uncertainty, Kant effectively reduced moral action to doing one's duty. At the same time, Kant provides a powerful rationale for adopting a less restrictive—and more ecosophic—understanding of our ethical obligations and opportunities. Finally, we will attend to the writings of Aldo Leopold, who first developed an ethics from ecological sensibilities.

Socrates and the Origins of Ethics

Socrates, the ancient Greek philosopher, might well be considered the Western world's first ethicist. Before he was condemned to death by drinking hemlock in 399 B.C.E., he had made something of a career of investigating the nature of ethical life. The origin of his vocation as a moral philosopher is legendary. His intellectual odyssey began in the footsteps of the wise men of his day. These philosophical types—today we might call them natural scientists—were the first cosmologists. They asked questions about the ultimate nature of reality and inquired after the fundamental building blocks of the universe. These queries were answered with a great deal of imagination and insight. The young Socrates, like his teachers, carried out such cosmological studies in his native city of Athens, the metropolis of the Mediterranean. Then one day a friend passed on some hearsay from a nearby town. The philosopher's life—and the discipline of philosophy—would never be the same.

The friend had visited the oracle at Delphi, a kind of soothsayer at a holy shrine a few days travel from Athens. There he learned that the oracle had made a strange pronouncement, observing that no man was wiser than Socrates. Upon hearing the story, Socrates was incredulous. But he was unwilling to dismiss the oracle's pronouncement out of hand. A fair trial was in order. Knowing that knowledge advances by way of the testing of hypotheses, Socrates set out to prove the oracle wrong. He was determined to meet with the wisest men of the land and probe their knowledge in conversation. As soon as he discovered a sage who knew more than he did, the Delphic prophecy would be disproved. Then he could go back to his studies of the cosmos.

Life has a way of happening while you are making other plans. Socrates's conversational efforts to disprove the oracle had unintended consequences. To Socrates's great surprise, neither the wise men he encountered nor the most prominent citizens of Athens could demonstrate firm foundations for any of their claims to knowledge. In the end, Socrates had to admit his own wisdom, but only because—unlike the other reputed sages and dignitaries—he knew one thing that they did not: he knew that he knew nothing. The beginning of wisdom, we now say with a bow to this ancient philosopher, is the admission of ignorance.

Socrates could refute all the learned men he encountered because, philosophically speaking, everything is hitched. All knowledge is interdependent. Just as every conclusion to an argument depends on the soundness of its premises, so every claim to knowledge is grounded on some yet-to-be-articulated conviction. Whenever a reputed wise man would make an assertion, Socrates would demand a justification of it. To supply the justification, Socrates's interlocutor would have to make other assertions. But Socrates would not allow these claims to be piled up one on the other unquestioned. Rather, he would thoroughly challenge each new assertion. Socrates was like a child persistently repeating the simple question, "Why?" every time her parent makes a demand and, in the face of resistance, attempts to defend it. This never-ending process of exposing the (shaky) foundations of all claims to knowledge is the Socratic method, sometimes called the dialectic inquiry. Adopting this approach to inquiry demonstrates that knowledge is not an island, entire of itself. It is a web.

Nuggets of knowledge that we take for granted are like individual nodes on a web. The node exists only because multiple strands cross each other and are in turn secured to other nodes. Our firmest convictions gain our self-assured endorsement only because of a confluence of other beliefs. No conviction, or piece of data, exists on its own without the support of multiple contiguous assumptions. And each of these assumptions, like filaments on a web, is itself secured by way of neighboring strands. While spider webs terminate on solid anchors, such as tree limbs or house walls, the web of human knowledge goes on indefinitely. The hitching never stops. The interdependence of knowledge claims means that we can never reach firm intellectual foundations. Like Indra's net, the web of knowledge has no terminus.

Effectively Socrates asked the, "And then what?" question in reverse. Rather than discerning the ramifications of every action or assertion, he investigated the chain of beliefs and values that brought about the action or assertion. Whenever someone stated a conviction, Socrates would explore the network of prior beliefs and assertions that provided the conviction with legitimacy. The Socratic method of inquiry, then as now, is an exercise in intellectual ecosophy. It demonstrates the pervasiveness of interdependence in the intellectual *oikos*. All efforts to leap off the web of knowledge in the misguided hope of planting one's feet firmly on the ground of certainty are shunned.

How does Socratic philosophizing, which effectively pushes one into endless intellectual inquiry, foster the development of an ethics? It is not obvious that it does. Certainly it was not obvious to the fifth-century Athenians who shared their city-state with Socrates. After a lifetime of badgering his compatriots with questions, the philosopher was accused, tried, convicted, and executed for impiety and corrupting youth. Ancient Athenians had severe doubts that philosophical inquiry would inculcate moral values. Religion, custom, and deference to authority seemed a more secure route.

Socrates's dialectic method, effectively asking the, "And then what?" question in reverse, had two dangerous consequences. In undermining all claims to certainty, it provided those who were wont to act unethically an excuse for self-serving behavior. After all, if all we know is that we know nothing, then how can one argue persuasively against unethical action. Perhaps everything is permitted. Alternately, Socratic inquiry might have the same effect as an attack from a stingray. As his student Plato explained in the *Meno*, it could numb victims and prevent their taking any action. At times, such paralysis might forestall evil deeds. But it might also induce bystanding, and the injustice implicit in passivity.

Justice and the Philosophical Life

The moral life, Socrates insisted, is the life of justice. But what is justice? That is the question tackled in Plato's *The Republic*, the most famous dialogue written about Socrates. At the beginning of *The Republic*, Socrates encounters a venerable old man named Cephalus who is on his way to pay his respects to the gods. To all appearances, Cephalus is a virtuous, pious citizen leading a good life. When Socrates questions him, however, the old man cannot give a reasoned defense of his actions or beliefs. To the extent that he acts justly and piously, he does so out of fear of the wrath of the gods. In placating the deities with offerings, Cephalus is hedging his bets. His piousness and justice are simply a means to avoid divine retribution in this world or the next. Cephalus's ostensibly virtuous actions amount to a form of bartering. The old man is not so much practicing virtue as engaging in commerce with the gods.

Socrates challenges Cephalus, and the old man soon quits the troubling dialogue to carry on with his daily routine. Socrates then hypothesizes that the truly just life cannot be engaged for some ulterior purpose, be it placating the gods, escaping punishment, or gaining the approval of one's peers. True virtue must be its own reward. Challenging this claim, Glaucon, one of Socrates's younger interlocutors, tells the story of Gyges.

Gyges was a poor shepherd in the service of the king of Lydia. One day, while Gyges was herding his sheep, an earthquake struck. The quake created an opening in the earth, and Gyges left his flock on the hillside to explore the cavern. Once inside, he came across a hollow bronze horse that contained a corpse-like body bearing a gold ring. Gyges stole the ring and made his way topside. Soon he discovered that the ring had magical powers. When he rotated it on his finger, he became invisible. The poor shepherd lost no time exploiting his magic. By making himself disappear at will, he could satisfy his every desire with impunity, gaining access to women, wealth, and power with no threat of being discovered or punished. After engaging in rape, thievery, and murder, Gyges eventually claimed the throne of the land. He became an all-powerful ruler, a tyrant.

Glaucon recounts this tale to make a point: if there were no punishment or retribution to be suffered from evil deeds, no one could resist the benefits of acting unjustly. Socrates disputes this claim. But he admits that the story of Gyges sharpens the debate. Assume for a moment that you have Gyges' magical ring. What would you do? How would you behave if you could commit any deed whatsoever without ever being discovered or punished? Temptation would be great. Indeed, the only person who could resist the spoils of injustice in the absence of any and all repercussions, Socrates suggests, would be the person convinced that justice is wholly its own reward. Only someone who valued justice for its own sake, regardless of what other benefits it produced, would act the same way with or without such magical powers. The story of Gyges' ring provides the ultimate test of our moral convictions.

Socrates spends much of the rest of the dialogue arguing that justice is indeed its own reward. As the just person may not be treated justly, Socrates must convince his interlocutors that it is better to suffer injustice than to commit it. This proved a hard sell, particularly in ancient Greece where taking good care of friends and severely punishing enemies was the standard definition of justice. Pacifism was quite unknown. The Christian dictum of turning the other cheek to one's assailant would not be articulated for another four centuries. For the ancient Greeks, suffering harm with no chance for revenge was the worst of fates—the fate of the slaves that Athenians exploited ruthlessly. Consequently Socrates was not very successful. By the end of the dialogue, even the most steadfast devotees of the philosopher remained unconvinced that justice is wholly its own reward.

It is not surprising that Socrates failed to win the day by reasoned argument. After all, he had taught his students that every claim in an argument could be thrown into doubt by inquiring after its justification. With reasoned argument unable firmly to settle the debate, Socrates resorts to telling a story. *The Republic* concludes with the Myth of Er, a tall tale about reincarnation.

At death, Socrates suggests, we are given the choice of what type of life we wish to lead when we return to the earth for another round. People who did not learn the lesson that virtue is its own reward while alive also tend to make bad choices when they are dead. Typically these recently deceased, unenlightened individuals choose lives that will give them access to the things that they craved in their previous incarnation. Such people often opt to be reborn into lives of wealth, honor, or power. But happiness will elude them again, as it eludes all who seek it in external goods. After leading wealthy, famous, or powerful but ultimately unfulfilling lives, they will return to Hades on their deaths to make new choices and engage in another round. After many such reincarnations, a few lucky souls will eventually learn their lesson and choose to be reborn into a life of justice. They embrace virtue for its own sake.

The fortunate few who figure out that virtue is its own reward will likely become philosophers. They do so for two reasons. First, only the philosophical life is pure, in the sense that only philosophers do what they do with no ulterior purposes or motives in mind. Doctors want to heal their patients, and practice medicine accordingly. But doctors also want to be well paid for their services. So doctoring mixes the art of healing with the art of moneymaking. The same is true for all other professions. The problem is not simply that professionals are remunerated for their services. More troublesome, they are often very concerned about their reputation. Whether to secure financial success or to satisfy their professional vanity, doctors, scientists and teachers may become more interested in appearing knowledgeable than being knowledgeable. The motive of moneymaking, securing a fine reputation, gaining social approval, authority, or fame competes with and frequently takes precedence over professional duties and responsibilities. The politician or ruler is no different. Although he may aspire to a life of justice, the effort to build a reputation, secure an office, and maintain authority exacts a cost. Power corrupts.

For all who practice a trade, the desire for wealth, honor, or power intrudes on the pursuit of the trade for its own sake. Unlike the sophists, the professional teachers of his day, Socrates was never paid for his services. He offered vexing conversation for free. The true philosopher, as his title indicates, is a lover (*philo*) of wisdom (*sophia*). As such, he is constantly seeking knowledge because it constantly eludes him. He incessantly puts the knowledge claims that present themselves in conversation to the test. Such philosophical activity proves self-fulfilling. If presented with Gyges' ring, the true philosopher would not alter his behavior at all. He would continue to converse, question, and test knowledge. If political power were thrust on him, Socrates observes, the philosopher would reluctantly accept the responsibility to rule as a means of repaying the community that raised and educated him. He would fulfill his responsibilities. But he would have no taste for power or its trappings of wealth and honor.

There is another reason that those who pursue virtue as its own reward choose the philosophical life. Virtue is a matter of desiring, and seeking to attain, not merely what we think is good but what really is good. No one who truly knew the good, Socrates claims, would fail to choose it. It follows that no one ever knowingly acts viciously. Vice is a form of ignorance. And virtue, Socrates concluded, is a form of knowledge. Justice is knowledge of what is right and wrong, and what each of us rightfully owes to others and is due from them. Courage is knowledge of when, where, and why to risk one's life or limb. Temperance or moderation is knowledge of what is excessive or deficient and how to steer between these extremes. If we know these things to be good, Socrates maintained, then we will surely pursue them. We will be just, courageous, and moderate. If virtue is a form of knowledge, however, it follows that those who want to be virtuous ought to pursue knowledge above all. They ought to become philosophers.

So the philosopher, and the philosopher alone, leads the pure life of virtue by way of his consistent and constant quest for knowledge, unblemished by the pursuit of money, honor, or power. The philosophical life is completely self-fulfilling. Neither the possession of Gyges' ring nor the changing circumstances of life would cause the philosopher to alter his behavior. Socrates provides the practical model for this claim. He is unwilling to prepare a speech for his trial, for example, and will not appeal to the emotions of his jury. Rather, he speaks in his defense to a capital charge before a wary, if not hostile, jury as if it were just another day in the *agora*, the public square, where he routinely engaged anyone he met with pesky questions. Effectively Socrates refuses to give up the philosophical life to become his own lawyer—even for a day. Jurors, who were used to being flattered by defendants, instead were forced to investigate—and asked to discard—their indefensible convictions, customs, and prejudices. The majority of the jury was not impressed with Socrates's harangue and found him guilty as charged.

But the philosopher was undaunted. While in jail awaiting the arrival of the lethal hemlock, Socrates refused to bemoan his woes, plan (or consider) an escape, or in any way alter his customary behavior or demeanor, despite pleas and protestations from visiting friends. Here, in the last hours of his life, Socrates continued to philosophize, tolerating only visitors who agreed to engage in inquiry and debate. Apparently when you have found something that is truly its own reward, neither the vicissitudes of life nor the certainty of death can throw you off track.

We have explored the Socratic ideal at length because it marks the first effort in Western thought to define the ethical life. Virtuous action is an end in itself, Socrates insists, never simply a means toward another end. To act morally or virtuously—being just, or courageous, or moderate—is its own reward. But the amazing strength and conviction that allows one steadfastly to live the life of virtue without being led astray by the lust for wealth, reputation, or power ultimately is grounded in the dismantling of all claims to firm knowledge. Ethical life can be resilient, Socrates demonstrates, but only for those willing and able to navigate seas of uncertainty.

Aristotle and Practical Wisdom

Focused as he was on pursuing knowledge, Socrates never bothered to write anything down. He left that task to his student Plato. What has been said of Socrates therefore might equally be said of Plato. Indeed, it is difficult to determine when Plato was faithfully recording Socrates's words and actions and when he employed his mentor's biography as a means to articulate his own views. In any case, Plato spent his life writing and teaching philosophy. He worked out of a garden park north of Athens called the Academy, which now serves as our word for an institution of higher learning. Plato had many students in his Academy, the most famous of whom was Aristotle. Much devoted to his teacher, Aristotle nonetheless disagreed with him on a number of important issues. Two such issues were Plato's effort to reduce virtue to knowledge and his accompanying effort to reduce dynamic community to a static unity.

Knowledge is not virtue, Aristotle insisted, because we often know what the right thing to do is, but fail to do it. He called this common experience *akrasia*, often translated as incontinence or weakness of will. *Akrasia* occurs whenever we act against our better judgments. Knowledge is not virtue, Aristotle held, because a virtuous character is primarily formed through habits rather than (formal) education. Moral virtue is acquired by way of practice. We become courageous by performing acts of courage, just as we become kind by performing acts of kindness. Many people, Aristotle observes, do not appreciate the impotence of academic learning in the matter of ethics. Instead, they prefer "theory to practice under the impression that arguing about morals proves them to be philosophers, and that in this way they will turn out to be fine characters. Herein they resemble invalids, who listen carefully to all the doctor says but do not carry out a single one of his orders. The bodies of such people will never respond to treatment—nor will the souls of such 'philosophers.'"[5] Talking and reading is no substitute for concrete experience and deeds. Only practice makes perfect—or at least starts one on the way to a virtuous life.

With a bow to Plato and Socrates, Aristotle acknowledges that the pursuit of wisdom is to be celebrated. But the philosophical life of contemplation, while "self-justifying," is not "self-sustaining."[6] To secure the conditions that allow the pursuit of knowledge—the basic needs of food, clothing, shelter, and security, as well as a community that provides education, training in the arts, friendship, and various amenities—one needs a more practical sort of knowledge coupled with the right sort of desire. Absent practical know-how and correct desire established through habit, Aristotle insisted, theoretical knowledge proves rather powerless. It will not automatically lead to virtue, because one may know one thing but desire and do something else. And it cannot secure the material and social conditions required for its own pursuit and development.

Aristotle employed the term *phronesis,* generally translated as practical wisdom or prudence, for practical know-how coupled with correct desire. Practical wisdom allows us to understand specific actions or events within a larger web of relationships. *Phronesis* grapples with the parts while understanding their place within and effect on the whole. It is attentive to the particular and sensitive to context. At the same time, it is informed by principle. It assesses the concrete without losing sight of the ideal. Aristotle's man of practical wisdom, the *phronimos,* employs his intellectual powers to determine what is good for the individual and the community. But the *phronimos* goes beyond identifying the components of a good life. Motivated by correct desire, he also pursues the good life and is skilled in his pursuit.

Phronesis is not simply knowledge; it is the capacity and disposition to put knowledge into practice. The *phronimos* exercises rather than simply comprehends the virtuous life while securing rather than simply identifying its worldly requirements. *Phronesis* melds practice with principle, stimulating one to act virtuously in concrete situations by integrating and coordinating the various parts of a good life into a well-balanced whole.

To be happy, Aristotle maintained, one must be able to exercise a full range of virtues. Practical wisdom facilitates their development. Indeed, to exercise any virtue requires *phronesis.* One can be courageous, for example, only by employing practical wisdom to determine which dangers ought to be faced, to what extent, at what time, and for what purpose. Putting oneself in harm's way without good reason is not courage; it is the vice of foolhardiness. Every virtue, Aristotle stipulates, is a mean between excess and deficiency. Practical wisdom determines where this mean is to be found at any particular time and place. In this manner, practical wisdom stimulates the development and regulates the exercise of all the other virtues.

Though habituated to virtuous action, the *phronimos* is not set in his ways. Quite the opposite is true. The practically wise person acts flexibly, continuously adapting his behavior to the situation at hand. He does not simply apply a principle or rule to determine what is right or virtuous. Neither does he merely follow custom or habit. Action is informed by thought, and, as importantly, thought is informed by action. So the *phronimos* engages in reflective deliberation that is attentive to circumstance. Aristotle thought so highly of the human capacity for practical wisdom that he declared the adaptive action of the *phronimos*, rather than an abstract, unchanging rule, to be the true "standard and yardstick" of justice.[7]

Aristotle argued that we should not attempt to fashion and regulate our ethical lives with inflexible tools demanding precise measurement and unbending conformity. Not the rigid carpenter's square but the "leaden rule" employed by the architects and masons on the island of Lesbos, Aristotle observes, is the appropriate model for moral living. A leaden rule can measure things better than a rigid square because it can bend to conform to various shapes. Likewise, our ethical conduct needs to be flexible and adaptive, the product of practical wisdom. That is because moral virtue, for Aristotle, is a matter of feeling and acting "at the right times on the right occasions towards the right people for the right motive and in the right way."[8] It follows that equity, understood as the bending and amending of the (written) law by way of practical wisdom, constitutes "the highest form of justice."[9]

Procrustes was a mythical robber who mutilated his innocent victims. He forced them to lie in a bed and would stretch or cut off their legs to make for a neat fit. An ethics reduced wholly to principles or rules will inevitably be Procrustean: it will achieve uniformity and consistency, but at the cost of violent standardization. Likewise, justice carried out too rigidly comes at a great price. The ancient Romans used to say *summum jus summa injuria*, meaning the greatest justice is the greatest injustice. Justice carried out too strictly becomes its opposite. As the poet Alexander Pope observed, "right, too rigid, harden[s] into wrong."[10] Owing to the diversity and uniqueness of individuals and the variability and unpredictability of circumstances, any effort to achieve an ideal by enforcing unchanging and unyielding rules will produce many injustices. The attempt to do just one thing, even if it is the quest for absolute justice, produces unintended consequences. The unyielding pursuit of something absolute means abandoning more moderate goals. In such situations, Aristotle states, the best becomes the enemy of the good.

Aristotle's notion of practical wisdom is not blind to the crucial need for rules, principles, and laws. Indeed, it relies on such standards as the starting points for inquiry and critical reflection. Absent these guideposts, the *phronimos* would be at a loss to determine what is equitable. But guideposts, although helpful in marking the way, do not determine the destination.

The superiority of practical judgment to static rules, principles, and laws follows from the nature of the human condition. Aristotle insists that "the *data* of human behavior simply will not be reduced to uniformity."[11] With this in mind, he holds that "ethics admits of no exactitude. . . . Those who are following some line of conduct are forced in every collocation of circumstances to think out for themselves what is suited to these circumstances."[12] The human condition is characterized by plurality. What Aristotle says of the political state applies as well to any community. It is made up of different sorts of people. A community of human beings is a diverse assembly of unique individuals (rather than, say, a uniform group of clones). It cannot be reduced to something singular, static, invariable, or homogeneous. A human community is a whole made up of distinct yet interdependent parts. Its web of ethical relationships is complex and dynamic, and it cannot be well navigated by invoking a rulebook. To be ethically responsible is to be socially responsive. One must adapt.

Morality, Duty, and Law

Two millennia after Socrates, Plato, and Aristotle walked the streets of Athens, Immanuel Kant attempted to develop an ethics that would also meet the test of time. Living in the Prussian city of Koenigsberg (and like Socrates, seldom venturing beyond his native town), Kant developed a moral philosophy without peer. He wrote in dense German prose, and about a good deal more than ethics. His moral philosophy, however, stands to this day as an exemplar of brilliant insight coupled with cogent argumentation.

Not unlike Socrates, Kant insisted that morality is grounded in rationality (as opposed to, say, custom or emotion). To be rational, for Kant, is to be an autonomous and intelligent agent capable of determining the rules that bind one's own conduct. These rules or laws set out moral duties—the things that reason tells us to do even when momentary desires (e.g., for pleasure or revenge) might push us in other directions.

In basing his moral philosophy on duty, Kant became the primary exponent of what came to be known as *deontology,* or duty-based ethics.[13]

What gives an action moral worth, for Kant, is never its prospective outcome, but the motive behind it. More specifically, an action gains moral worth when it is motivated by duty, that is, by our will to abide by a law that we, as rational agents, have made for ourselves to govern our conduct. It is not simply conformity to this law that produces moral worth, but our acting for the sake of the law. Socrates insisted that virtue is to be exercised for its own sake and is its own reward. Likewise, Kant insisted that duty is to be done for its own sake, not for what it produces or achieves. Virtue, for Kant, is the moral strength of will to fulfill one's duties.

The only thing that is good in itself, according to Kant, is a good will. A good will is wholly governed by a sense of duty to the moral law. And what is the moral law? It is a principle of reason that governs action unconditionally, regardless of circumstances, consequences, desires, or preferred outcomes. The primary and fundamental moral law, from which all other obligations are generated, is the categorical imperative.

The categorical imperative, Kant's most famous invention, states our unconditional obligation to always "act only according to that maxim by which you can at the same time will that it should become a universal law." In other words, reason tells us to act in such a way that were every other rational creature (that is, all other people) to follow our example, the conditions that make such actions possible would not be undermined. An example will help. The maxim "Always tell the truth" could be consistently adopted by everyone. The maxim "Tell a lie whenever doing so best serves your interests" could not. Following the latter maxim might allow one to achieve a particular outcome in a particular circumstance, perhaps avoiding punishment for a misdeed by denying one's guilt. But if everyone adopted this maxim, our trust in speech would vanish. Giving one's word would be a meaningless gesture, for we would all be aware that self-serving lies are perfectly acceptable and frequently employed. In such a world, lying would seldom, if ever, serve our purposes. It would not help us avoid punishment for a misdeed because no one would believe our protestations of innocence. The maxim to lie whenever we want could never become a universal law without undermining the conditions of its own potency.

According to Kant, then, we have a duty always to tell the truth. Do we also have a duty to protect the environment, pursue social justice, and be economically farsighted? Do we have an obligation not to undermine the prospects of future generations? For the deontologically inclined, Kant provides a foundation for such convictions. Kant's categorical imperative, one might say, formalizes the obligation to live sustainably. We have a duty to live in such a way that everyone could follow our example without the (ecological, social, and economic) conditions that make such lives possible being undermined. In other words, we have a duty to sustain the community of life that sustains us.

Moral life, for Kant, requires acting with a steadfast will in strict accordance with rules that might be universally adopted. The community of life, however, is dynamic and evolving. It cannot be sustained by instituting a rigid, rule-based plan. Sustaining it demands practical wisdom and adaptive practices. While the categorical imperative might be employed in support of sustainability, Kant's deontological morality bears the same pitfalls as Plato's pursuit of justice through static roles and obligations. To explore how Kantian ideals might be adapted to support an ethic of sustainability, we will visit with the first person to craft moral principles from ecological considerations.

Community, Practical Wisdom, and Principled Action

Aldo Leopold may well have had Aristotle in mind when he wrote, "All ethics so far evolved rest upon a single premise: that the individual is a member of a community of interdependent parts."[14] Actions are ethically right, Leopold states, when they preserve the integrity and stability of the community; they are wrong otherwise. Our sense of obligation and ethical responsibility, Leopold observed, arises from the loyalty and affection we feel toward other members of our community.

For Aristotle, the community in question was quite restricted. It was limited to citizens of the city-state or *polis*, though in some respects it was extended to all Greeks. The Stoics (Greek and Roman philosophers who followed Aristotle in developing ethical theory) proposed extending the moral community to include all human beings, regardless of which *polis* or country they inhabited. They called themselves cosmopolitans—citizens of the *cosmos* or universe. Our contemporary understanding of universal human rights finds its origins with these early cosmopolitan thinkers.

Leopold did the Stoics one better. The Stoics extended ethical care and moral responsibility across geographical space, beyond the confines of the city-state. Leopold extended ethical care and moral responsibility across

species, beyond the confines of the human race. He proposed the development of a "land ethic" that "enlarges the boundaries of the community to
include soils, waters, plants, and animals, or collectively: the land."[15] What
Leopold says of "land," of course, applies equally to the skies and seas. His
land ethic includes everything that participates in the "biotic community,"
which includes humans, plants, animals, and the various ecosystems they
inhabit.[16]

The poet John Donne famously wrote, "No man is an island, entire of
itself. . . . Any man's death diminishes me, because I am involved in
mankind; and therefore never send to know for whom the bell tolls; it tolls
for thee." What Donne declared for men, Leopold declared for other
species. Humanity is not an island, entire of itself. It is one of many species
participating in the community of life. Within this broadest of communities, the extinction of one diminishes all. Keeping "every cog and wheel"
in the biotic apparatus is not only a requirement of prudential tinkering.
For Leopold, it is a moral obligation. The bells of extinction, which toll
incessantly today, make ethical claims. We cannot afford deaf ears, Leopold
counseled, and we should be well beyond inquiring for whom the
bells toll.

What would it mean to extend ethical concern to the biotic community,
as Leopold recommends? We know that it forbids us to throw away any
of the parts. But this does not mean that we cease using, and at times
consuming, certain parts. After all, the community of life is largely defined
by relationships of competition, predation, and consumption. Absent
these relationships, ecosystems would collapse. In the midst of our participating in the dance of life, however, Leopold states that we should always
act to ensure the integrity and stability—and, he adds, the beauty—of the
biotic community. To this end, we must rely on practical wisdom because
no universal, unchanging rules can be applied. The biotic no less than the
social *oikos* is a diverse and dynamic entity. It takes practical wisdom to
determine, at any particular time and place, what strengthens the stability,
integrity, and beauty of the communities to which we belong.

With this in mind, Leopold observed, as Aristotle did, that ethics are
never written in stone. Life is too diverse and dynamic to allow static rules
and laws to capture it once and for all. Rather, ethics are "tentative" products of "social evolution." They are tentative, Leopold states, "because
evolution never stops."[17] Like Aristotelian ethics, Leopoldian ethics are

oriented to the welfare of plural, changing communities composed of interdependent participants. While principled and visionary, such an ethics is attentive to context and adaptive.

Practical wisdom goes beyond static rules and principles to determine what is ethically appropriate and required in any particular circumstance. But as we noted, guideposts are always helpful. A brief trip back to Koenigsberg, and subsequently to the oldest moral traditions of humankind, may help set down useful guideposts for a Leopoldian ethics intent on preserving and enhancing the stability, integrity, and beauty of the community of life.

Kant provided a number of versions of the categorical imperative. Exploring one of these versions reveals how Kantian ethics effectively obligates us never to treat ourselves or other members of our community as merely one thing. Consider this formulation: "Act so that you treat humanity, whether in your own person or in that of another, always as an end and never as a means only."[18] Rational beings should always be treated with "respect," Kant argued, because they have autonomy and are capable of self-directed behavior. Human beings, as rational creatures who can legislate their own lives, have moral worth, or dignity. As such, they should never be reduced merely to an instrument serving another person's needs or wants. To treat someone solely as a means is to treat a being with inherent dignity as if it were a mere tool. Slavery is the epitome of immorality, in this respect, as the slave is by definition an instrument serving the master's whims. But slavery is not the only case where the moral respect due to individuals is denied. Kant deemed any action immoral that even momentarily or partially denies or neglects the dignity and intrinsic worth of another human being.

When I pick up a hammer to bang in a nail, I treat the hammer wholly instrumentally. It is one thing and one thing only to me: a tool to achieve my explicit purpose of pounding a nail. I need not consider the hammer otherwise. When I interact with another human being, Kant insists, the situation is very different. I must not treat her solely as an instrument to be used for my benefit. Although I may interact with her for some particular and even self-interested purpose, I must not lose sight of her moral worth. I may approach the grocer to buy vegetables, hail the taxi driver to gain transport, or cooperate with a fellow worker or teammate to achieve a common goal, but in the midst of such self-interested, purposeful action,

I must always treat other individuals with the respect due to fellow rational creatures.

While necessity dictates that we interact with people to gain and share services and goods, morality dictates that we also recognize people as (potential) cocreators of a common world. That is what it means for human beings to be self-legislators. To live ethically, for Kant, one must always respect people as autonomous creatures who, like oneself, are capable of directing their own lives. They should never merely be providers of services or goods. We must also, at all times, encounter them as ends in themselves.

Kant's categorical imperative, particularly in the latter version just discussed, provides a formalized restatement of one of the oldest, and most widely adopted, moral laws known to humankind: the golden rule. Thales, the ancient Greek sage who predates Socrates by more than a century, is reputed to have said: "Avoid doing what you would blame others for doing." Some version of this ethic of reciprocity is found in virtually every religious and moral tradition. The sacred text of Hinduism, the Mahabharata, claims that the sum of all duty is contained in the dictum. "Do naught unto others which would cause you pain if done to you."[19] The Udana-Varga of the Buddhist tradition states, "As it is the same thing that is dear to you and to others, hurt not others with what pains yourself."[20] In the Analects, Confucius writes, "Do not impose on others what you yourself do not desire."[21] When challenged to teach the holy book with brevity, Hillel, the Jewish sage, declared, "What is hateful to you, do not do to your neighbor, that is the whole Torah, while the rest is the commentary thereof."[22] Likewise, the Christian gospels stipulate that the maxim, "So always treat others as you would like them to treat you," sums up all previous moral instruction and law.[23]

The golden rule can be interpreted as a personal morality, dictating how an individual should regulate her interactions with others. It may also be interpreted as a social ethic, dictating how a community should organize its collective life. If a community were to implement the golden rule, social policies would be crafted to ensure justice, where each member of the community, including the least wealthy or powerful, receives his or her due (something we all want for ourselves).

Living by the golden rule requires treating others justly, with the understanding that the best guide as to what justice to others entails in any given circumstance is the sort of treatment you yourself would want to receive. But doing unto others as you would have done unto you does not mean acting as if everyone shares your personal likes and dislikes. It does not mean serving only pistachio ice cream to guests because ice cream is your favorite dessert and pistachio is your favorite flavor. That might be appropriate if you lived among clones. But we do not live in homogeneous communities where everyone's likes, dislikes, hopes, aspirations, and projects are the same. In diverse, plural communities, the golden rule requires treating each person with the respect due to a unique individual capable of self-direction. It requires treating each person as a cocreator of the community.

Living by the golden rule is a stretch for most of us, though certainly a comprehensible and viable ideal. But what would it mean, practically speaking, to apply the golden rule to all other members of the biotic community? How might we show respect for a creature we are eating or making into a pair of shoes? More generally, is it possible to live in moral relationship with all members of the community of life?

We Are All Related

Native dwellers across the globe lived according to a land ethic long before Leopold coined the term. They demonstrated a capacity for ecological respect. The Lakota Sioux, like many other aboriginal peoples, have an expression for this respect. They say *mitakuye oyasin* when sitting down to a meal, heading off on a hunt, entering a sweat lodge, or beginning a sacred ritual. *Mitakuye oyasin* means "all my relations" or "we are all related." The Sioux expression underlines that all creatures participate in and benefit from an interdependent community of life. It is an acknowledgment of ethical responsibility, emotional and spiritual connection, and existential gratitude. Saying *mitakuye oyasin* invites the spirits of all members of this community to participate in a sacred moment. Lakota, like many other native peoples of the Americas and other lands, speak of the animals they rely on for food and clothing as their brothers and sisters. They are all relations.

Other ethical and religious traditions have developed similar sensibilities. A Ladakhi prayer from the region of Kashmir asks for the quick attainment of Buddhahood for all the animals used or eaten that day. Buddhists believe that all forms of life undergo repeated reincarnations until they have learned fundamental lessons that allow them to escape the cycle of death and rebirth (not unlike the cycle of reincarnation depicted in Socrates's myth of Er). The plants and animals that Buddhists use and eat are encountered as co-travelers of a grand, spiraling journey. Everything is hitched in time.

The interdependence of beings in cycles of life and death grounds the Buddhist ideal of the Bodhisattva. A saintly figure on the path to enlightenment, the Bodhisattva dedicates herself to the liberation of all beings from suffering. Her every word and deed is offered as a gift for the benefit of sentient life. As Frances Cook observes, the Buddhist ethic "reaches beyond the human, so that if I throw away a paper drinking cup, I can almost hear the reverberations of a falling tree. . . . Consequently, if I have the least shred of concern for my own spiritual progress, I must care—really care—for the spiritual growth of even the grass." Of course, we walk on grass, cut down trees, and eat daily that which once lived. We could not sustain ourselves otherwise. The Buddhist ethic does not deny this fact of life. It acknowledges that we must "destroy and consume in order to continue to exist." But in the face of this reality, the Buddhist takes life only when necessary to sustain life, and does so with respect and gratitude. To waste, and in particular to waste life, Cook observes in describing Buddhist morality, "is the rankest sort of ingratitude. It nullifies the thing we depend on, murders it, and in so doing, we murder ourselves and others. . . . Perhaps respect and gratitude are more important than we may think, ranking near the top of those modes of action we call ethics."[24] Compassion, respect, and gratitude for all life forms, whatever their individual paths to Buddhahood, is central to Buddhist ethics. This compassion, respect, and gratitude cannot be withheld—indeed, it is especially important to express—when we are using and consuming our fellow participants in the web of life.

A Tibetan practice illustrates how such compassion, respect, and gratitude can be cultivated. One is asked to imagine every living thing that one encounters, and particularly those things one consumes, as having been in another life the prior reincarnation of one's mother. The gratitude and affection one naturally feels for one's mother can then be directed to the tiger, bird, fish, goat, and spider. In this way, one may experience and cultivate a sense of connectedness to every member of the earthly community. This practice expands appreciation of the interdependence of all beings across space, with a host of diverse life forms that share the planet. It also expands appreciation of interdependence across time. Like many aboriginal traditions, Buddhism acknowledges every past, present, and future organism's participation in a sacred web of life.

An ethics of interdependence does not require a belief in reincarnation or in our ability to communicate with animal spirits. It does require that we acknowledge as cocreators of community those life forms that share our social and natural habitats. The aboriginal sense of relatedness, like the golden rule found in so many religious and ethical traditions, describes a morality of reciprocity, an ethic of caring for and respecting others as connected parts of a whole. Ecosophic awareness grounds these traditions.

We are parts of a greater whole and cannot assume to know how all the parts are connected. A sense of humility befitting this general state of ignorance is in order. Like Socrates, we ought to recognize that knowledge of our fundamental ignorance is the beginning of wisdom. Not knowing how all the parts fit together might then prompt us to treat each part, at some level, as an end in itself. When we use or consume a member of the community of life—a practice inseparable from membership within this community—compassion, respect, and gratitude are in order. This sensibility might take any number of forms. It may be implicit in our frugal use of resources. It may be explicit in prayers of thanks before meals. It may take the form of a silent or voiced tribute whenever we venture into the web of life to interact with fellow travelers: *mitakuye oyasin*.

The Place of Love

William James observed that "if your heart does not *want* a world of moral reality, your head will assuredly never make you believe in one."[25] Ethical life is aided by an intellectual appreciation of the breadth and depth of our interdependence with other members of our community. But it is grounded in our affection for them. The caring sense of connection that grounds ethical relationships at the most fundamental level is the experience of love. Love defers and attenuates our sense of individuality. The *me* does not cease to exist when one loves, but it becomes tentatively and temporarily displaced, put in abeyance, or backgrounded. An emotional

and spiritual connection to another develops that outshines one's attachment to the self and its interests.

For Aristotle, we recall, living ethically means living virtuously in pursuit
of the good life. Importantly, the good life cannot be achieved by the
individual alone, no matter how virtuous he or she is. It is only possible
for those fortunate enough to have friends. The Greek word for friendship
is *philia*, which is generally translated as love (the term also employed to
describe the philosopher's relationship to wisdom). Friendship, a relationship of love, provides the ideal condition for the practice of virtue and
reflects the consummation of such practice.

The core of ethical life, Socrates and Plato state, is justice. But, Aristotle
observes, community is grounded not only on justice but on friendship.
Justice demands that we satisfy our obligations to other members of our
community. Friendship requires not only that individuals be given what
is due to them but what is good for them. When we love a person, we
want the best for him or her. Justice is grounded in mutual obligation.
Friendly love is grounded in a mutual concern for the greatest well-being
of another individual.[26] For Aristotle, a human community requires justice
to survive but friendship to flourish. Beyond a willingness to meet our
responsibilities, the desire to pursue the good of others, to seek their greatest well-being—however circumscribed in daily affairs and however spo-
radic in its fulfillment—is crucial to the welfare of a community.

The virtuous person, Aristotle states, is above all a friend to himself.
Indeed, he interacts with friends as if they were second selves. Friends
provide and extend opportunities to love, exercise virtue, gain pleasure
from observing virtue exercised by others, and experience gratitude for
existence. Aristotle maintained that one should love one's friends as oneself
and love one's self as a friend. A life enriched by such love is the most
fertile field for the full development of virtue.

The golden rule that holds pride of place in Judeo-Christian ethics is
often accompanied by the dictum to "love your neighbor as yourself."
Originally found in the Jewish scriptural text Leviticus (19:17–18) and
reiterated in the Christian gospels, the dictum to love your neighbor as
yourself is perhaps the greatest ethical commitment imaginable. It certainly demands more than the Kantian edict to treat the self and others as
ends rather than merely means. Treating a person as an end, Kant
stipulated, meant treating her with the respect due to an autonomous,

self-legislating, rational creature. To love someone requires respecting an individual's rational, self-legislating capacities. But it demands more. One must also acknowledge and encourage the development and exercise of all his or her worthy capacities. To love is to embrace the full person and to embrace a person's full potential. St. Bernard said that to love an individual is to want that individual to be—not simply because it serves one's purposes, as one might want a hammer to be if there is a nail to be pounded. Love means wanting something to flourish for its own sake.

In community, the sensibility of interdependence begins to break down the rigid boundaries of the egoistic individual. An ethics grounded in justice extends this process. The further dissolution of these boundaries at an emotional level is achieved by love. In an ethical community grounded in justice, we understand that we rely on each other for our sustenance and welfare and have responsibilities to the members of the community on which we rely. The welfare of others becomes of interest to us and presents us with obligations. With love, we experience—perhaps in a fleeting and ineffable fashion—another's welfare as our own. We experience another's happiness *as* happiness.

Those who practice justice still need friendship to complete their lives, according to Aristotle. But friendly love exceeds the boundaries of justice. Friends, Aristotle insists, do not need justice to regulate their relationships. Love, it is often said, knows no rules. That makes love—as anyone who has loved well knows—a dangerous affair. Playing by the rules is always safer. But rules never capture what is most evolved in the realm of ethics. As William James wrote, "The highest ethical life . . . consists at all times in the breaking of rules which have grown too narrow for the actual case."[27] Practical wisdom pursues the highest justice by going beyond enforcing rules to achieve equity. Likewise, love pursues the greatest good for others by going beyond mutual obligations to achieve the fullest sense of connection.

That love knows no rules and exceeds the boundaries of justice demonstrates both its power and its danger. Social communities, unlike a small group of friends, cannot exist without rules. Love is too personal a force to provide the regularity and accountability required for large organizations. Here justice is required. But justice is not, for that reason, the most developed or best form of human relationship. With this in mind, Rumi, the thirteenth-century Persian poet, asks us to move beyond justice, beyond doing what is right and not doing what is wrong. He wants to meet us in a place beyond obligations.

Loving friends find themselves in a place beyond obligations, a place of love. For most communities, however, establishing and maintaining fields where love can grow requires the protective boundaries of justice. We do not require a sense of justice to cultivate love for family and friends. Within the broader community, however, the path of reciprocity and responsibility, the path of justice, is the most secure route to greater connection. Justice is the glue that binds together the social fabric such that friendship and love might take root.

In social communities, the golden rule of reciprocity and forms of justice that highlight relationships of interdependence are pathways to the love of neighbor. The neighbors in question need not be limited to other human beings. Other species may be included. Sociobiologist E. O. Wilson theorizes that human beings developed a love of nature or *biophilia* as the product of living within interdependent natural communities for millennia.[28] We came to love that which gave us life. However our love of natural things originated, it provides today, along with our love of fellow human beings, a crucial ingredient of ethical community. "It is inconceivable to me that an ethical relation to land can exist without love," Aldo Leopold wrote.[29] The same may be said of our ethical relationships with human co-travelers.

Ethics is the practice of becoming increasingly attuned to and responsible for the effects and side effects of our actions and nonactions in an interdependent community. Walking this path of justice prepares us to enter the field of love. Here, at the apogee of ethical life, self and other are, paradoxically, both dissolved into a greater whole and cherished for their uniqueness.

From Theory to Practice

Pursuing justice, developing virtue, following the golden rule, acknowledging all our relations, and loving our neighbors, human and otherwise: it all sounds wonderful—in theory. But what does it mean in practice? How do we translate this ethical vision into practice, moving from aspiration and ideal to the actual and concrete? At the end of the day, might it simply be wishful thinking? Or worse, might it, like many other well-intentioned plans, have pernicious unintended consequences?

Aristotle has been a helpful guide in our explorations of ethical theory. Like all other guides, however, he came with his own baggage. Aristotle defended slavery and relegated women to second-class status as citizens. Relying on Aristotle may have turned off readers who reject his severe moral prejudices. Likewise, invoking Garrett Hardin in the previous chapter may have produced similar side effects. Although Hardin is a champion of the notion of interdependence, he is antagonistic to any morality grounded in love or altruism. He endorses "lifeboat ethics," legitimating sacrifice of the lives and livelihoods of poor people in order to secure resources for the rich. We need not defend a thinker or actor as a whole to benefit from something insightful or useful he has said or done. Just as we may beneficially employ Aristotle's understanding of practical wisdom without endorsing slavery, so we may embrace Hardin's ecological insights without adopting lifeboat ethics or other features of his moral theory. At the same time, Hardin's rejection of humanitarian approaches to ecological problems presents a formidable intellectual and moral challenge. With this in mind, I will take on the task of outlining how an ethics of interdependence might be translated into practice by critically evaluating Hardin's claims.

Moral theorists occasionally explain an individual's concern for the welfare of a broader community as the extension of affective relationships that naturally form between kith and kin. Edmund Burke, the eighteenth-century Irish political theorist, wrote, "To be attached to the subdivision, to love the little platoon we belong to in society, is the first principle (the germ as it were) of public affections. It is the first link in the series by which we proceed towards a love of our country and of humankind."[30] The natural care we have for the near and dear, Burke suggests, will eventually extend itself to a broader community. Yet loving one's own platoon does not always evolve into a more encompassing embrace. A brief survey of the killing fields of history suggests that fear and hate of those existing outside one's group is commonplace. Indeed, this fear and hate may often provide the centripetal force required to hold an inner circle of brethren together. It is said that a nation is a group of people with nothing in common except an enemy. The comment is cynical, but much history is on its side.

Evolutionary biologists have argued that the moral glue that binds a community together is not derived from affection or a sense of duty. Rather, moral norms are generated in response to external threats, as a survival mechanism. Richard Alexander writes that "rudimentary moral systems (indirect reciprocity) will appear where outside threats most powerfully dictate group cohesion, when such threats are combated best by complex social organization within the group, and when the actions of single individuals or small subgroups can threaten, from within, either the group as a whole or its most powerful elements."[31] Garrett Hardin concurs that the threat to welfare posed by an out-group provides the cohesive force of reciprocity and love that knits together an in-group. Hardin writes that "if we desire a world in which altruism can persist we must reject the ideal of One World and consciously seek to retain a world of more or less separate, more or less antagonistic units called (most generally) tribes. . . . Altruism toward brothers—reciprocal altruism *among* brothers—must be coupled with a measure of antipathy toward others."[32] If Hardin's assertion is correct, this is a harsh truth indeed. It does not bode well for a world of growing interdependence and grave dangers.

Hardin believes that antipathy toward others is required because we face scarce resources. He asks us (the citizens of rich nations) to imagine ourselves sitting in lifeboats that float on seas virtually awash in (poor) people clamoring to be saved. Should we invite the drowning masses aboard our small craft? The Christian ideal of being your brother's keeper, Hardin observes, would advise as much. And just as surely, it would produce an overloaded craft, with deadly consequences. Hardin tersely writes of the side effects of a morality of loving your neighbor as yourself: "The boat swamps, everyone drowns. Complete justice, complete catastrophe."[33] Without some antipathy toward others—or at least a much-diminished sense of responsibility and care—we might allow the drowning masses entry to our lifeboat, ensuring tragic results. Self-interest must trump altruism to ensure survival for the fortunate few. Hardin concludes: "It is unlikely that civilization and dignity can survive everywhere; but better in a few places than in none. Fortunate minorities must act as the trustees of a civilization that is threatened by uninformed good intentions."[34]

Responding to Hardin's ethical theory and policy recommendation, Barry Commoner writes: "Here, only faintly masked, is barbarism."[35] Commoner spoke for people within and beyond the environmental community in voicing his outrage. But Hardin would deny the charge of barbarism, claiming realism as his principle. Indeed, Hardin develops his own "cardinal rule" of realism: "Never ask a person to act against his own self-interest." Abiding by this rule, Hardin admits, sets severe limitations on our individual and collective endeavors. But ignoring the cardinal rule is at best quixotic. At worst, Hardin predicts, it will endanger civilization.

Hardin provides an example of what happens when you neglect the cardinal rule of realism. Environmental activists from the Greenpeace organization were on the high seas trying to stop the killing of whales. The activists, at times thrusting themselves heroically between the harpoons and the whales, plaintively addressed their combatants. Hardin writes:

We are told of idealists on board this [Greenpeace] vessel who appealed by megaphone to the captain of a Russian whaler to cease his activities in the interests of the whales and posterity. The captain's reply was, of course, of the sort that we of the older generation call "unprintable." And why should it not be? Whatever sneaking admiration we may have for the idealists of the Greenpeace Foundation—and I confess I have more than a little—their program is quixotic because it violates the Cardinal Rule by asking people to act against their own self-interest.[36]

The cursing captain did not stop killing whales in the face of the protests. But were the Greenpeace activists really tilting at windmills in their courageous mission?

In fact, the antiwhaling campaign mounted by Greenpeace (along with other groups, such as the Sea Shepherd Society), has proven quite successful. It has not completely stopped the hunting of whales, but it has raised public awareness across the globe, altered national policies, and stimulated an internationally monitored moratorium on most forms of whaling subscribed to by most countries, including Russia (where whaling is now restricted to that carried out by the indigenous Chukchi and Yup'iit peoples in the Siberian territory of Chukotka). Greenpeace has been successful, but not because it won over sea-bound whalers. Rather, Greenpeace skillfully exploited modern media to inform and inspire the general public and persuade their political representatives. These appeals were not to narrow self-interest. Greenpeace explicitly asked people to violate Hardin's cardinal rule by considering first and foremost the welfare of future generations and the welfare of the whales themselves. And for the most part, it worked.

The cardinal rule and lifeboat ethics might appear the appropriate conclusion to an argument grounded on the premises of evolutionary biology. If survival of the fittest demands the unflagging pursuit of self-interest and if this form of egoism can be expanded beyond the individual to include the welfare of others only when external threats dictate the strategic benefits of solidarity, then Hardin is on pretty solid ground. Human beings are certainly no strangers to self-interest, and in many instances tribalism (strategic solidarity) is clearly the product of external threats. However, both the pages of history and our personal lives testify to the existence of motivations beyond narrow self-interest and fear.

Even from an evolutionary perspective, the pursuit of self-interest necessarily leads the individual beyond narrow egoism. Despite the most calculated efforts at self-preservation, our bodies will inevitably disintegrate in a few score years. Despite the accumulation of vast personal wealth, not a penny will make it past our graves. And despite the greatest feats of self-propagation, our genes will dissipate over the years to the point of being statistically negligible. Because only half our genes get passed on with each succeeding generation, our great-great-great-grandchildren will bear little more genetic similarity to us than they do to contemporary strangers living on the other side of the globe. Self-interested pursuits are limited inevitably to short-term success. The self is a short-term affair.

In contrast, the legacy of our contributions to community and to moral life may endure for millennia. The golden rule and Aristotle's concept of practical wisdom have been around for thousands of years and are still strong. Kant's categorical imperative continues to exert its influence after many centuries. Indeed, every contribution to the welfare and sustainability of one's community has the potential to outlive and outshine the self by millennia.

Hardin was an advocate of systems thinking, yet he failed to realize, as another proponent observed, that "the systems approach begins when first you see the world through the eyes of another."[37] Thinking like a mountain expands one's perspective in space and time. In a world of limited, finite selves, systems thinking encourages us to see and act beyond the blinders of short-term self-interest. Systems thinking has ethical implications.

Hardin's hard-nosed realism is not based on fancy. It is grounded on the prevalence of narrowly self-interested behavior among human beings in a world of scarce resources. As we cannot deny this reality, we should attend to its consequences. But the meaning of self-interest is not static. Global ecological, technological, economic, political, psychological, and physical interdependencies are growing. Securing long-term self-interest in such a world—for individuals and nations, rich and poor alike—is best achieved by prudential cooperation. The community that sustains us and that we have a self-interested responsibility to sustain no longer ends at familial, local, regional, or national borders. Climate change is only the latest and most pressing of the many phenomena that highlight global interdependence. We do not live in a world of self-enclosed, easily defended lifeboats. Rather, we sit atop a global seesaw that is currently overloaded to the breaking point and gyrates quite out of balance, threatening to topple all from their perches. Those who sit at one end of the creaking and swiveling board cannot unilaterally control their fate. And marching down to the other end of the seesaw to throw off competitors would only worsen the situation. Cooperation guided by foresight is demanded.

In proposing the "first law of human ecology," Hardin was taking aim at naive moral and political idealists who ignore the cardinal rule. The "ecological interaction of things," Hardin observed, "constantly surprises us and negates our laboriously worked out plans for reform."[38] Hardin provides many fine examples of how well-intentioned reforms proved disastrous because their proposers failed to learn the lesson that you can never do merely one thing. The sixteenth-century Florentine political theorist, Niccolò Machiavelli, knew as much. Often dubbed the first political realist, Machiavelli wrote:

It seems that in all actions of men, besides the general difficulties of carrying them to a successful issue, the good is accompanied by some special evil, and so closely allied to it that it would seem impossible to achieve the one without encountering the other. This is evident in all human affairs, and therefore the good is achieved with difficulty, unless we are so aided by Fortune that she overcomes by her power the natural and ordinary difficulties.[39]

The good commonly accompanies the bad. In a world of unintended consequences, fine motivations often meet with foul results.

Saving the lifeboat of civilization will not be easy, and it will not be achieved by any one party acting unilaterally even if charitably. Within the aboriginal traditions of Australia, it is said that "if you have come to help me, you are wasting your time. But if you have come because your liberation is bound up with mine, then let us work together." In an age of interdependence, unilateral altruism is no more a strategy for success than avarice. We cannot discharge our ethical obligations to human beings across the globe without the difficult work of entering into interactive relationships with them.

Unintended consequences result from all human action, not only that spun out of idealistic plans for reform, as Hardin suggests. Naive do-goodism that neglects the first law of ecology serves no one in the long term. But people acting in their narrow self-interest are not exempt from this law. They too cannot do just one thing. Self-interested behavior has side effects. These cannot be calculated or controlled, and in the long term, they often undermine the explicit goals of self-interested actors. History is filled with bloody tales of people and nations that were undone by their untrammeled quest for gain or glory. The blinkered pursuit of self-interest is an enduring reality, but it is not a recipe for success.

Hardin's own principles suggest as much. If he is correct that self-interest is king and its rule is seldom, if ever, contravened, then it follows that the vast majority of the social and ecological problems we face today are the product of self-interested behavior. That should be reason enough to question the long-term benefits of people doggedly pursuing their narrow self-interest. Ignoring the first law of ecology is always asking for trouble, even if the one thing that you are trying to do is defend your lifeboat. This is especially true in a world of seesaws.

Conclusion

Humans develop ethics to regulate their interactions within communities. Moral obligations and rules of reciprocity are intended to secure individual well-being as well as the health, welfare, and sustainability of the greater community. To think and act ethically entails viewing oneself and others as interrelated parts of a greater whole. Ethics concerns the responsibility to sustain the community that sustains us.

Sustainability, in this respect, is inherently ethical. It is a commitment to well-being that has community—most expansively, the community of life—as its foundation. As an ethical commitment, sustainability directs itself to the welfare of current and future generations, with the understanding that those who follow us have the right to inherit a world in which the natural environment, equitable social structures, avenues for learning and cultural creativity, and economic opportunities have been maintained or improved.

Human desire is inexhaustible. The world is not. Living ethically helps us achieve happiness even in the absence of the ring of Gyges. As important, it helps us to avoid the mayhem that the unlimited pursuit of desire produces. In an interdependent world, every action and event—from our morning meal to late-night conversations—has ethical implications and should be engaged with consciousness. Clearly ethical life is not easy. It demands the pursuit of justice and the development of a wide range of virtues, including the virtue of practical wisdom. Yet living ethically allows us to flourish and gain fulfillment in a finite world. In contrast, the blinkered pursuit of self-interest ultimately is a futile quest. It cannot sustain the self or the world.

As members of the community of life, we depend on each other for our lives and livelihoods. Such relations of interdependence make ethics both possible and necessary. What our ethical responsibilities are, however, can never be decided in the abstract or with finality. They can be discovered only for particular individuals or societies in particular circumstances, based on their roles in the community that sustains them. Practical wisdom grounded in experience allows this discovery. Practical wisdom is systems thinking applied to our ethical *oikos*.

Thinking like a mountain—rather than an isolated individual—is an ethical as well as an ecological exercise. Leopold might have had Ralph Waldo Emerson's understanding of the natural world in mind when he developed the land ethic. In his famous essay "Nature," Emerson wrote:

Nothing in nature is exhausted in its first use. When a thing has served an end to the uttermost, it is wholly new for an ulterior service. . . . Every natural process is a version of a moral sentence. The moral law lies at the centre of nature and radiates to the circumference. It is the pith and marrow of every substance, every relation, and every process. All things with which we deal, preach to us.[40]

If nature preaches a moral lesson, it mostly concerns the uniquely human conceit that we can do just one thing. Even actions that serve intended ends to the utmost extent are wholly new for ulterior service. Interdependence and the law of unintended consequences are the compelling themes of nature's sermon.

It is difficult, if not impossible, to be responsive to this homily when one's purview is hemmed in by egoism. Pursuing narrow, short-term self-interest produces too many unintended consequences. One such side effect is the reinforcement of the tendency to think and act in a narrow, short-sighted, self-interested manner. Just as one becomes courageous by carrying out acts of courage and kind by practicing kindness, so one becomes increasingly self-interested and narrow-minded by carrying out acts of narrow-minded self-interest. This effect is predictable, if mostly unintended. Its impact on the self and the world is dire.

For Aristotle, living ethically required acting justly in the world to fulfill one's obligations, establishing and strengthening personal habits so as to make the development and exercise of virtue both more likely and more complete, and engaging in self-fulfilling acts of moral excellence. Each of our actions has the potential to have these multiple reinforcing effects. Negative synergies, of course, are also possible, and inaction is not without consequence. To be practically wise is to know that you can never do merely one thing while also knowing that passivity is not a viable option. On the web of life, every act has effects and side effects, including the act of bystanding. Frances Moore Lappé writes:

Since interdependence isn't a nice wish, *it is what is*, there can be no single action, isolated and contained. All actions create ripples—not just downward through hierarchical flows but outward globally through webs of connectedness. And we never know what those ripples might be. Beneath our awareness, perhaps, we are coming to realize that our acts do matter, all of them, everywhere, all the time. . . . [E]very choice we make sends out ripples, even if we're not consciously choosing. *So the choice we have is not whether, but only how, we change the world.*[41]

Practical wisdom prompts us to address the question, "And then what?" before taking any action. At the same time, it marshals the moral courage to act. Acting responsibly in the face of uncertainty is *the* ethical challenge. We are all world changers. The question is whether we can become wise, just, and loving ones.

A story about an old Jewish rabbi nicely underlines the challenges in meeting our moral responsibilities within a worldly web of relations. A usually temperate and kind-hearted cobbler had libeled his neighbor, a grocer. He spread gossip that painted the merchant in a very unflattering way. The cobbler came to regret his unkind words but could not bring himself to face his victim, so he sought counsel from the village rabbi, who lived up the hill. The rabbi heard the cobbler's story and sensed his remorse. "Take a pillow from your bedroom," the rabbi said. "Go to your rooftop, rip the pillow apart, and throw the feathers into the wind. Then come back to see me in three days." The man was befuddled by the advice, but dutifully returned to his house, climbed to his roof pillow in hand, and carried out the strange request.

Three days later, he walked back up the hill to see the revered teacher. "I did exactly as you suggested, Rabbi" the cobbler reported. "Is all forgiven now?" "Not quite," said the rabbi. "Go now and gather all the feathers, and bring them to me." The cobbler was aghast. "That is impossible," he remonstrated. "They have been scattered to the four winds. There is no telling where they might be by now." The rabbi looked at him sternly yet kindly. "Indeed," he said. "And so it is with your unkind words. They can never be retrieved, and who knows what evil they continue to do. Now go and make amends to the grocer."

The rabbi was practically wise and a builder of community. Every action, both virtuous and vicious, enters a web of relationships. Its effects ripple out in every direction. As daunting as this moral reality may be, it is also empowering. The beneficial effects of our virtuous endeavors also may resound indefinitely. For better and worse, our deeds constitute our legacy to life. The ethics of interdependence celebrates this fate and challenges us to make every act count.

3 Technology

In short, we can say today that man is far too clever to be able to survive without wisdom.
—E. F. Schumacher

In Greek mythology, a titan named Prometheus was given the job of distributing to all the beasts of creation their unique capacities. For reasons unclear, Prometheus allowed his absent-minded brother Epimetheus to carry out the task. Epimetheus took on his new charge with gusto. He gave the lion its powerful jaws and claws, the bear its brawn, the birds their feathered wings, and the snake its venom. At the end of a very long day, Epimetheus was greatly satisfied with his work—until he realized that he had left one species, humankind, empty-handed. On learning of his forgetful brother's blunder, Prometheus was much chagrined. Because he was unwilling to leave the human species defenseless in the world, Prometheus returned to Mount Olympus, the home of the gods, to secure something of merit. He lit a torch from the sacred fire of the sun and, stealthily placing a burning ember in a fennel stalk, slipped back down the mountain to bestow the purloined gift to the deprived race.

In short order, the all-powerful Zeus, king of the gods, discovered the thievery and duly punished Prometheus. Nevertheless, humanity now had what it needed to defend and distinguish itself. Fire allowed human beings to frighten off their predators, cook their meals, warm their homes, harden their pottery, forge their metals, and communicate with each other over long distances. Humankind was now unique among earthly creatures by its tremendous power to refashion its world according to its own needs, wants, and visions. Technology had arrived on the young earth.

Zeus, however, saw fit to punish the human species for receiving the theft of fire. He had a beautiful woman created to become the wife of Epimetheus, and subsequently presented the happy couple with a marvelous jar (some say a box) as a wedding present. Expecting retribution, Prometheus (whose name means *foresight*) had warned his brother to beware any gifts from the gods. But Epimetheus, good to his name (which means *afterthought*), ignored the advice. The beautiful woman was Pandora. When the newlyweds opened Zeus's enchanting gift, all hell broke loose. The jar had been filled with every vice imaginable. Pandora, watching with horror as greed, envy, vanity, hate, and a host of other evils were loosed on the world, quickly shut the lid. But it was too late. The damage had been done. The only thing remaining in the jar was hope, which now would stay sealed away forever.

In presenting his version of this well-known tale in the *Theogony*, the Greek poet Hesiod concluded that it is impossible to hide from the mind of Zeus.[1] Perhaps the ancient bard was simply admonishing his readers not to accept stolen gifts from an all-seeing, and rather vengeful, god. Or perhaps Hesiod was saying that from a cosmic perspective, everything is connected. Although humans may well hanker after power, it always comes with a price. Every technological good is accompanied by an unexpected consequence that will be loosed on the world. Prometheus's gift was the first technological benefit to exact, in Zeus's revenge, its unknown cost. The human race would experience countless more. It would profit greatly from the metals forged in red-hot hearths, for instance. But this great boon would make warfare more deadly and more likely, as primitive clubs of wood gave way to swords and spears, cannon and rifle, bomb and missile.

Hesiod was admonishing us to ask, "And then what?" before we act. With similar concerns, Wendell Berry, the agrarian man of letters, counsels us "to worry about the predominance of the supposition, in a time of great technological power, that humans either know enough already, or can learn enough soon enough, to foresee and forestall any bad consequences of their use of that power."[2] The marvelous wits that allow for our Promethean craft prove too meager to foresee, avoid, or remedy every unintended consequence. Upstream ingenuity will always exceed the ability to control downstream effects.

Barry Commoner first identified this problematic aspect of technology in the early 1970s. The subtitle of Commoner's highly acclaimed book, *The Closing Circle*, was "Nature, Man and Technology." Commoner's work raised concerns about the social and ecological effects of a centralized, technological way of life while demonstrating the predominant role that technology has played in fostering environmental and social problems. Readers of *The Closing Circle* could no longer naively revel in the promise of technology. They had to be concerned about its potential mischief.

Commoner argued that technology, rather than overpopulation, has caused most of the world's pollution and that problems of distribution, not global shortages of food, have caused most of the world's famines. The methods and products of industrial production and distribution, he said, are largely to blame for environmental woes. Industrial technology increasingly takes us beyond organic cycles and closed-loop systems. With the introduction of synthetics, and even more so synthetic toxins and radioactive material, recycling was made more difficult, if not impossible. Terminal disposal became the only option for "goods" that reached the end of their usefulness, and safe forms of disposal remain in short supply.

For these reasons, Commoner argued, technological efforts to improve on nature are always dangerous and generally detrimental. The tools, machines, chemicals, substances, and processes that we invent are often promoted as liberating solutions to age-old problems, but when it comes to technology, Commoner insisted, there is no such thing as a free lunch: every technological benefit leaves a debt to be paid.

Technology is wonderful. When it is developed wisely and selected well, it may contribute greatly to the welfare of a particular *oikos* or community. But the unintended consequences of technological endeavors cannot be escaped; they can only be mitigated. And every increase in technological power is accompanied by an increase in risk.

Technological Counterproductivity, Side Effects, and Misuse

Technological devices and processes produce side effects. But often their direct effects are most harmful. That is, technology often proves counterproductive, contributing to the problem it was designed to solve. The proposed technological cure proves worse than the disease.

Consider the valiant effort of the United Nations Children's Fund (UNICEF) and other international agencies to provide clean drinking water in Bangladesh. Until the 1970s, the majority of rural Bangladeshis relied on surface ponds and rivers as their main sources of water. Uncontrolled sewage contaminated much of these surface waters, unleashing a battery of water-borne diseases. It was estimated that a quarter of a million children died each year from bacteria-contaminated drinking water. UNICEF joined with the World Bank to fix the problem. The agencies initiated the drilling of a vast number of tube wells—long, narrow pipes placed in deep holes—that would tap into the country's underground aquifer, where pollution from sewage would be absent or negligible.

Although the water from the tube wells proved largely free of bacteria, about half of them—as many as 10 million wells—were contaminated by arsenic, a naturally occurring but highly toxic chemical element widespread within the geological formations of the region. As people started developing tumors and dying from arsenic poisoning, officials began testing specific wells to determine if they were contaminated. To date, however, only a small fraction of the millions of potentially lethal tube wells have been tested. It will take years, if not decades, for the problem to be adequately addressed, assuming that the resources and political will needed to do so are generated. Meanwhile, as many as 70 million Bangladeshis may now be drinking water from wells containing lethal levels of arsenic. Tens of thousands of them will likely die from the toxic water each year in what is called the "biggest mass poisoning in history"[3] Agencies keen on improving Bangladesh's access to clean drinking water could have taken the obvious measure of testing tube wells for toxicity as they were being dug, but they did not. Officials trying to solve one problem, bacteria contamination of surface water, had latched onto a technological solution, tube wells, without asking, "And then what?" Good intentions provide no immunity from the first law of ecology.

Another example of a counterproductive technology involves a gasoline additive designed to reduce pollution. Beginning in 1979, methyl tertiary butyl ether (MTBE), which is derived from natural gas and petroleum, was added to motor fuel in the United States. MTBE, an oxygenate, raises the octane level of gasoline, which helps it burn more efficiently, reducing tailpipe emissions of benzene, carbon monoxide, and volatile organic compounds. Previously ethyl, a lead-based oxygenate, had been used for this purpose. But concerns about the toxic effects of lead eventually forced it to be taken off the market. MTBE seemed a healthy alternative to leaded gasoline.

The production and use of MTBE in the United States increased by almost 5,000 percent over the two decades following its introduction. Beginning in the 1990s, this growth was largely a product of efforts to cheaply satisfy oxygenate requirements of the Clean Air Act. By the end of the decade, more than 200,000 barrels of MTBE were being manufactured every day. MTBE was a quick-fix success story: it reduced tailpipe emissions without adding toxins, and the air was better for it. But MTBE's effect on water was another story.

Gasoline stations typically store their fuel in underground tanks. When these tanks leak, as many do, gasoline and its additives are released into groundwater supplies. MTBE is highly soluble in water and gives it an unpleasant taste. Beginning in the mid-1990s, thousands of water wells and aquifers across the country became contaminated by MTBE. Regulatory agencies were caught unawares.[4] In response to widespread concerns, California and New York banned MTBE in 2004, with many other states following suit in subsequent years. By 2006, most gasoline companies began phasing out MTBE, often using ethanol as a replacement. Hundreds of unresolved lawsuits are still pending regarding MTBE contamination of private and public drinking water. The cost of cleaning up these groundwater sources is estimated to be many billions of dollars. Effectively, MTBE cured one disease by creating another. The health of the patient did not improve.

Most of technology's unintended consequences are not strictly counterproductive. The technology achieves its intended purpose well enough. But in doing so, it creates a raft of problems that may outweigh the benefit received. Consider the case of chlorofluorocarbons (CFCs).

Though invented in the late 1800s, CFCs were first effectively synthesized in the 1920s and quickly became widely used as refrigerants. Subsequently they were also employed as cleaning solvents and propellants. Decades after CFCs became a global market success, it was discovered that the chemical had two extremely dire side effects. When the largely inert gas was released into the air, it did not break down. Rather, the chemical slowly migrated up to the stratosphere. Here it established its second and third careers—not as refrigerant, solvent, or propellant but as a destroyer of ozone molecules and a potent greenhouse gas.

Ozone shields the biosphere from the harmful effects of the sun's ultra-violet radiation. Without stratospheric ozone, life as we know it would not exist on earth. The thinning of ozone caused by CFCs allows higher levels of ultraviolet light to penetrate to the planet's surface. The result is higher rates of skin cancer, eye cataracts, and immune deficiencies among humans. It also adversely affects terrestrial and aquatic plants and animals. In turn, CFCs are second only to carbon dioxide among greenhouse gases in terms of their contribution to climate change.

International protocols were developed to limit the production of CFCs and other ozone-depleting gases beginning in the late 1970s. Notwith-standing the success of these global agreements, ozone depletion will continue to worsen for decades as previously released CFCs systematically work their way up to the stratosphere, where they destroy ozone for many years before becoming inactive.[5] Indeed, it will be the middle of this century before the stratospheric ozone layer begins to recuperate in earnest. Four score years after the development of CFCs, much organic life across the planet remains its victim. We will suffer from its unintended conse-quences for decades to come.

The Green Revolution provides another case of a technology that deliv-ered on its promise but created pernicious side effects. Beginning in the 1940s and reaching its height in the 1970s, the Green Revolution was created through agricultural technology. High-yielding, disease-resistant crops such as wheat were developed by cross-breeding to increase produc-tivity. The goal was to feed the quickly growing populations of developing nations.

The Green Revolution is said to have spared 1 billion people from hunger and starvation. There is no denying this tremendous benefit. But the side effects are equally patent. Across the globe, aquifers have been dangerously depleted because high-yielding crops demand vast amounts of irrigation. In turn, fertilizer and pesticide use, which monoculture agri-culture requires, has increased massively. Only a small portion of the nitrogen supplied by artificial fertilizer, about a third for grain crops, actu-ally finds its way into agricultural products. The rest leeches out into air and water supplies, where it proves pernicious. And pesticides, as we have seen, have their own side effects. The Green Revolution also stimulated the growth of large-scale agribusiness, abetting the concentration of power and wealth and its accompanying social inequities. Finally, some argue that overpopulation in developing nations was much exacerbated by the Green Revolution's impact on food production.

In a speech commemorating the thirtieth anniversary of his 1970 Nobel Peace Prize, American agronomist Norman Borlaug (1914–2009), the father of the Green Revolution, announced that agricultural technology would again change the world. He predicted that genetic engineering would double current food production, a tremendous feat that, he said, would be required to feed the earth's growing population "on a sustainable basis."[6] Critics argue that this second green revolution will prove no more sustainable than the first. And they have some good research to buttress their charge. A study of the use of genetically engineered (GE) crops in the United States from 1996 to 2009 demonstrates that more herbicides were used than would have been required had non-GE crops been planted. The biotech plants had been genetically engineered to tolerate certain herbicides, but the increased use of these herbicides led to the increased growth of herbicide-resistant "superweeds."[7] A vicious circle was born.

Consider a third example of well-intentioned technology that had untoward consequences despite achieving its intended purpose: the Aswan High Dam in Egypt. The dam, a wondrous feat of engineering, was built on the Nile River in the 1960s to control flooding, produce electricity, and increase water available for agriculture. It met these goals admirably and was widely acclaimed as a boon for Egypt's poor and growing population. To be sure, 90,000 people were displaced by the 300-mile-long and 35-mile-wide Lake Nasser that was created when the Nile was plugged. But flooding was controlled, electricity was produced, and farms gained a steady supply of water for irrigation. Of course, there were side effects.

For starters, more than 6 feet of water (over 14 cubic kilometers in total) evaporate off the surface of Lake Nasser each year, which means there is much less water flowing down the Nile and reaching the Mediterranean Sea. In turn, the creation of the dam has halted the downstream delivery of more than 130 million tons of silt and soil a year. This silt and soil that the river carries from Ethiopia formerly supplied the Nile valley with nutrients to replenish those lost to agriculture. It now settles on the bottom of Lake Nasser. Owing to the loss of nutrients delivered by floodwaters, agricultural production along the lower Nile must be sustained through fertilizer use, at a high monetary cost to farmers and an ecological cost to the farmlands and local water supplies.

Deprived of much of the silt from the Nile, the ecosystem of the eastern Mediterranean Sea has also been adversely affected. The sardine population has virtually disappeared. The nitrates and phosphates in the silt that used to wash into the sea produced phytoplankton blooms. The main food source of the sardines was the zooplankton that consumed this phyto-plankton. Absent silt from the Nile, this ecological web collapsed. In turn, the year-round irrigation now necessary along the river in the absence of annual flooding has increased the salinity of the soil. Agricultural produc-tion levels on these increasingly salty fields has already decreased by as much as 50 percent. The Nile valley was farmed continuously for 5,000 years between periodic floods. Now irrigated and flood proof, it may have to be forever abandoned within the century because of salinization. Mean-while, Egypt's population has risen from under 30 million at the time of the Aswan Dam's construction to over 80 million today, in some part owing to the dam's initial benefits.[8]

Now consider a technology that has proven both counterproductive and generative of damaging side effects: biofuels. Biofuels are derived from biomass, which is to say, from recently grown organic material. Burning biofuels produces carbon dioxide. However, the harvested plants providing the biomass recently sequestered their carbon from the atmosphere. If more plants are grown to replace the harvested plants, these new plants will sequester even more carbon. In theory, then, burning biofuels need not add to the net amount of carbon in the atmosphere. With this carbon equilibrium waved as a banner, biofuels have been promoted as a sustain-able substitute for fossil fuels and a remedy for global warming.

The most widely used biofuel worldwide is ethanol, a form of alcohol produced from the fermentation of starches and sugars derived from food crops such as corn, wheat, sugar beets, and sugarcane. Given the way most of these crops are grown and processed, biofuel is far from carbon neutral. The problem is that the production of biofuel today is highly dependent on fossil fuels. When one adds up all the carbon dioxide released into the atmosphere from the use of farm machinery, from the production and application of fertilizers, pesticides, herbicides, and fungicides made from fossil fuels, from the heat-drying of the corn and the ethanol fermentation and distilling processes, and from transportation, the numbers are damning.

In turn, the fields of biomass used to produce biofuel release large amounts of nitrous oxide as a product of natural growth processes. Nitrous oxide is hundreds of times more powerful per molecule as a greenhouse gas than carbon dioxide is. On the basis of nitrous oxide releases alone, some scientists assert, biofuels contribute more to global warming than does the continued use of fossil fuels.[9]

To make matters worse, forests, peatlands, and grasslands are often converted to cropland or plantations to produce biomass fuels. When this is done, organic material (trees, peat, and grasses) that previously sequestered carbon no longer provides this service. As a result, depending on the biomass cultivated and the type of land converted for its cultivation, the use of biofuels can increase the production of greenhouse gases by over 30 percent of that which would occur through extracting, refining, and burning equivalent amounts of fossil fuels.[10]

The conclusion is clear: as a hedge against global warming, ethanol and many other biofuels make no sense whatsoever. Importantly, the carbon footprint of biofuels is not its only negative impact. There is the increased threat to biodiversity caused by the conversion of species-rich forests into cropland. There is the water pollution caused by the fertilizers and biocides employed in crop production. There is the depletion of water resources caused by the irrigation of biofuel crops and the distilling of ethanol.[11] Finally, we must account for the global food shortages and price spikes caused by the use of grains for fuel that formerly supplied nutrition. In a world where 1 billion people already suffer from hunger and malnutrition and where food shortages not only make for human misery but also generate social unrest, refugees, and political instability, the use of food crops for biofuel may prove disastrous.[12] Although some biofuels, such as those produced from algae and switch grass, may prove to be carbon neutral and avoid some if not all of the related problems, much research and development is still required. In most cases today, biofuels are a problem masquerading as a solution.

The unintended consequences of technology occur when inventions generate harms that their inventors did not foresee. At times, these harms arise because the tools or processes invented for one purpose come to be employed for quite another. The history of dynamite is a classic case.

Swedish industrialist and engineer Alfred Nobel invented dynamite in 1866 by mixing nitroglycerin with silica. Mr. Nobel was constructing bridges and buildings in Stockholm, and he needed a powerful and reliable method for removing rock. Dynamite did the trick. Nobel made a great deal of money through the patenting of his new invention, but those who bought his "blasting powder" were not limited to contractors constructing bridges and buildings, carving out quarries, or mining ores. Before the end of the decade, dynamite was destroying men, not rock. By the late 1880s, the U.S. government had gotten into the "dynamite gun" business, building large (15-inch) pneumatic weapons that would fire dynamite-filled shells at their targets from up to a mile away. Nobel quite unintentionally had opened the gates to modern warfare.

Alarmed at the harm his invention was causing, the wealthy Nobel, in his last will and testament, left most of his inheritance to the establishment of annual prizes for individuals from any country who "have conferred the greatest benefit on mankind." Nobel died in 1896, a year after he wrote his will. Kin contested his posthumous philanthropy in the courts for a spell, but by 1901, the first Nobel laureates were named.

Along with awards in physics, chemistry, medicine, and literature, Nobel designated a prize for "the person who shall have done the most or the best work for fraternity between nations, for the abolition or reduction of standing armies and for the holding and promotion of peace congresses."[13] These prizes were meant to offset in some small way the deadly misuse of Nobel's own invention. Ironically, Nobel's prizes have honored many whose inventions proved every bit as deadly as Nobel's own. Nobel laureates have been honored for discoveries and inventions that in time contributed to the development of chemical warfare and military munitions (Fritz Haber and Carl Bosch) and the building of nuclear bombs (Ernest Rutherford, Niels Bohr, Enrico Fermi, Otto Hahn, and Harold Urey). Technology allows us to fashion amazing tools. It does not determine how these tools will be used or by whom.

Understanding Technology

Whether developed for profit or designed to serve altruistic purposes, technology creates risks, many of which cannot be anticipated. The significance of these risks increases in tandem with the scope of the technological endeavor. Given the impact of contemporary technology, thorough calculations of costs and benefits, risks and opportunities are needed. But due diligence does not grant us immunity from unintended consequences. However painstaking our calculations are, something will escape.

All human actions produce unintended consequences. So what is special about technology? On one hand, the answer is, "Nothing at all." The development and use of technology is like any other human endeavor: it sends a ripple across the web of life whose full ramifications can never be known. On the other hand, technology is unique in its relation to unintended consequences because it is specifically designed to do just one thing. That is its objective, its raison d'être. And this singular objective makes technology an imperial force. A brief return to ancient Greece will help explain this claim.

As the myth of Prometheus illustrates, technology made humankind special. It constituted a particular form of knowledge, which the Greeks called *techne*. *Techne* employs models, principles, or patterns in order to make things. Aristotle contrasted *techne* with *episteme* (science) and *sophia* (wisdom), human capacities that allow us to know things. He also contrasted it with *phronesis* or practical wisdom, which allows us to do things together. *Techne* is unique among these capacities because it is not an end in itself, as is the case for the pursuit of knowledge by *episteme*, theoretical contemplation grounded in *sophia*, or virtuous action grounded in *phronesis*. Rather, the sole end and purpose of *techne* is to bring a specific product into the world. It makes a useful thing that fulfills a specific purpose. The purpose may be transportation, as is the case when we make a wagon or spaceship; destruction, as is the case when we make a spear or nuclear bomb; or the pleasant occupation of our time, as is the case when we make puzzles, games, or films. In each instance, *techne* is oriented to making things according to a plan or design. *Techne* is always assessed in terms of its finished product.

Tool use is not wholly unique to our species. Chimpanzees use twigs to gather termites, and Egyptian vultures use rocks to break open ostrich eggs. Still, the human species has developed its technology to an extraordinary degree. The ability to employ complex conceptual models to produce food, clothing, shelter, security, tools, and mechanical devices sets the human species apart from all other creatures. The ancient Greeks celebrated this human capacity, and so should we. At the same time, *techne* is not the only form of knowledge, and technological activity is not the only way that humans can interact with each other and their world. Building on

Aristotle's thought, the German philosopher Martin Heidegger (1889–1976) examined how technology has colonized and often usurped the diverse ways humans can relate to their world.

Heidegger observed that humans are unique because of the distinct and variable ways in which they interact with their environment. The human species can relate to the world as if it were the object of pure knowledge. For this, the skill of *episteme*, or scientific inquiry, is required. We can also relate to the world poetically, as artists, by giving voice to the world's diversity and mystery and making it meaningful. We can relate to the world as an object of contemplation, employing *sophia*. And we can relate to the world as virtuous doers, employing *phronesis*.

Beginning with the ancient Greeks, Heidegger suggests, humanity's various ways of relating to the world became increasingly technological. More and more, the world was seen as something to be possessed, controlled, and used for specific purposes. The human species transformed the earth into its quarry. In turn, we extended this same instrumental way of interacting with things to people. In the vocabulary of the modern business corporation, our fellow men and women became *human resources*. In the contemporary world, Heidegger argues, the ancient skill of *techne*—one among many ways of being in the world—has gained a virtual monopoly on our modes of knowing and acting.

In Heidegger's vocabulary, technology transforms the earth and all its inhabitants into "standing reserve" (*Bestand*). Everything is effectively perceived as a resource, a raw material to be put to human use according to a plan. For the man equipped with only a hammer, everything is treated like a nail. For technological man, the natural and social world becomes a storehouse of resources awaiting exploitation. In the world of technology, Heidegger writes, "everywhere everything is ordered to stand by, to be immediately at hand, indeed to stand there just so that it may be on call for a further ordering."[14] Heidegger was concerned with the totalizing reach of a mode of human thinking and acting that always and everywhere reveals the world as standing reserve.

The development of a particular tool, machine, or technological process is not in itself a problem. Developing and using technological implements and artifacts can produce great benefits. Without the knowledge and skill of *techne*, we would find ourselves back in the Stone Age. Indeed, without *techne*, we would not be fully human. The capacity to shape the world according to visions and plans is a crucial feature of our species. The problem arises when a technological frame of mind becomes, effectively, the only frame of mind. At this point our tools and machines become the effects of an expansive technological drive that recognizes no boundaries and makes no distinctions in its appropriation of the world. Everything is put into technological service, including the earth's diverse life forms, with no exception made for the human species.

Marshal McLuhan, the Canadian communications theorist, observed, "All technology has the property of the Midas touch; whenever a society develops an extension of itself, all other functions of that society tend to be transmuted to accommodate that new form."[15] Heidegger would agree. Technology has the Midas touch, and a particularly contagious one at that. Everything that it comes in contact with turns into standing reserve. Whatever technological man interacts with becomes an extension of himself. His fingerprints and footprints are everywhere. Through the application of science and engineering, technology reaches across the planet, deep into space, within the cosmic forces of the most elementary particles, and into the genetic components of our being. Heidegger invoked the ancient Greek sage Heraclitus who said, "Nature loves to hide." Today there is nowhere to hide from technology's Midas touch. Formerly a vast and mysterious force, nature now lies exposed, and shaped, by human hands.

In his meditation on the "end of nature," the environmental writer Bill McKibben offers a lament in accord with Heidegger's concern. McKibben suggests that nature has been irreversibly integrated into humankind's utilitarian calculus, permanently brought out of hiding and altered. How, McKibben asks,

can there be a mystique of the rain now that every [acidic] drop—even the drops that fall as snow on the Arctic—bears the permanent stamp of man? Having lost its separateness, it loses its special power. Instead of being a category like God—something beyond our control—it is now a category like the defense budget or the minimum wage, a problem we must work out. . . . What will it mean to come across a rabbit in the woods once genetically engineered "rabbits" are widespread? Why would we have any more reverence or affection for such a rabbit than we would for a Coke bottle?

McKibben is forlorn by the demise of nature. But he offers a ray of hope:

Someday, man may figure out a method of conquering the stars, but at least for now when we look into the night sky, it is as Burroughs said: "We do not see ourselves reflected there—we are swept away from ourselves, and impressed with our own insignificance. . . ." The ancients, surrounded by wild and even hostile nature, took comfort in seeing the familiar above them—spoons and swords and nets. But we will need to train ourselves not to see those patterns. The comfort we need is inhuman.[16]

Heidegger would again concur. Yet it is ever more difficult to find inhuman comfort in a world so thoroughly transformed by technology.

Today light pollution from ever-illuminated cities makes stargazing nearly impossible for urban denizens. If we escape the metropolis to spend an evening in the wilds beholding the starry firmament, we cannot but be distracted by the scores of illuminated satellites that traverse the horizon. Multinational corporations have even explored the possibility of building huge Mylar billboards to advertise their wares from the heavens, as well as gigantic stratospheric reflectors to allow nighttime industry. Perhaps the starry firmament will not provide an "inhuman" comfort much longer. In turn, the genetic manipulation of plants and animals accelerates without pause. No less than the heavens above, the inner world of organic life is increasingly transformed by humans. With each passing day, the likelihood grows that as we look high into the sky, or deep within nature, the primary impressions that we receive will be those made by human hands.

E. J. Mishan observed that "while new technology is unrolling the carpet of increased choice before us by the foot, it is often simultaneously rolling it up behind us by the yard."[17] This Faustian bargain occurs in large part through the loss of local flavor, cultural diversity, and enduring craftsmanship that occurs when we submit to a steady stream of mass-produced, cheap, prepackaged, processed, and plastic consumables shipped to us from the far reaches of the globe. But in a more fundamental sense, it also occurs owing to the loss of nontechnological ways of being. The danger is that technological thinking and acting may severely erode—and perhaps destroy—traditions of intellectual inquiry, meditative reflection, aesthetic creation, moral responsibility, political action, and spiritual connection.

Technological endeavors aim at doing just one thing: the efficient production of useful artifacts or processes. Neither the side effects of the means employed to produce these things nor the side effects of loosing these technological creations on the world need come into consideration. The ability to act with such singular focus, Heidegger argues, arises from the ability to perceive the world with a singular focus. In other words, one will attempt to do just one thing only if one already sees the world as just one thing. Today, Heidegger suggests, we increasingly see the world as one thing: as standing reserve, a pool of resources awaiting exploitation. Given this singular way of perceiving the world, it stands to reason that technology will be developed and used with little thought to unintended consequences.

Notwithstanding the potentially devastating effects of technology, Heidegger does not counsel us to reject or abstain from the development or use of tools, machines, or engineered processes. We cannot retreat to some Pleistocene, pretechnological way of life. But neither should we resign ourselves to the imperial domination of technology. Technology has its place. It can meet human needs and help sustain the larger community of life. We have the option of sustainable technology.

Solving for Pattern

Wendell Berry insisted that when we act to solve a problem, we may have special ends in mind, but our actions do not have specialized effects. Although technology is specifically designed to do just one thing, it cannot achieve this singular purpose. In the face of this challenge, Berry coined the phrase "solving for pattern" to identify how sustainable technologies can and should be developed.

Bad technological solutions, owing to ignorance or disregard of the larger networks or patterns that support them, prove destructive of these patterns. As Berry writes, "A bad solution solves for a single purpose or goal, such as increased production. And it is typical of such solutions that they achieve stupendous increases in production at exorbitant biological and social costs."[18] Good solutions, in contrast, are attentive to and reinforcing of the relationships and networks that support them. Good solutions solve for pattern, addressing a particular need in a way that sustains the web of relations within which it is embedded. Solving for pattern turns the obstacle we face in never being able to do just one thing into an opportunity. In Berry's words, "A good solution solves more than one problem, and does not make new problems. . . . It is the nature of any organic pattern to be contained within a larger one. And so a good solution in one pattern preserves the integrity of the pattern that contains it."[19] Rather than attempting to solve a perceived problem with a technological quick fix—an attempt that inevitably produces unintended and generally pernicious consequences—Berry asks us to approach every problem as a symptom of a system that is failing to satisfy the needs of its constituents. Any attempt to satisfy these needs must simultaneously strengthen the network of relationships that characterize the broader system or community.

Consider the "problem" of periodic dull wits. The mind seldom operates at full capacity. For lack of focus or energy, we often find ourselves without the intellectual or creative power to get a job done well or quickly. Sufficient sleep and exercise are reliable means of sharpening wits and stimulating imagination. Drinking caffeinated beverages is another, albeit short-term, remedy for dull wits. And pharmacologists have developed other means of "turbocharging" the brain.[20] These chemicals increase cognitive focus and energy. They are quite effective, fast, and potentially long-lasting.

So we have a choice. Faced with the problem of dull wits, we might pharmacologically stimulate a portion of the brain. Alternatively, we might refresh and reinvigorate the entire mind and body with sufficient sleep and exercise. The former approach solves the problem of dull wits but will likely cause other problems. Individuals who choose pharmacological means to heighten and prolong intellectual performance, for instance, might forgo even the minimal levels of sleep and exercise needed to maintain their physical and emotional health. And of course there are potential medical side effects of the drugs. The latter approach, though not as quick, solves for pattern. It improves not only mental resilience but also physical and emotional well-being. It approaches our intellectual and creative capacities within the broader context of the "pattern," or web of relations, that we call the human mind and body.

The widespread use of drugs for the treatment of high cholesterol provides a similar lesson. Lipitor, produced by Pfizer, is currently the world's best-selling drug, with annual sales that reach well over $10 billion. Other cholesterol-lowering drugs, including other statins, are on the market, and new ones are being developed. The use of drugs to offset the cardiovascular effects of a diet rich in meats and dairy products and low in exercise may indeed decrease the risk of heart disease and heart attacks, but it may have medical side effects that exceed the benefits. And since such drugs treat only the symptoms rather than the causes of the problem—unhealthy eating and exercise habits—other health problems arising from the same habits, such as diabetes, may actually increase. And this is to say nothing of the heightened carbon footprint produced by diets high in meat and dairy and sedentary lifestyles.

These examples might give the impression that solving for pattern means opting out of technological solutions altogether (forgoing drugs for regular exercise, sufficient sleep, and healthy diets). But employing the right technology is often a crucial component of solving for pattern. An agricultural problem in need of a good solution provides an example.

Given badly eroded pastureland, how do you produce prodigious amounts of beef, pork, rabbit, turkey, chicken, and eggs? One option is the industrial model of agribusiness: raising and feeding animals in separate enclosures with harvested grain until they are big and fat enough to ship to distant markets, where they will be sold, butchered, and packaged for further transport to retail outlets. This entails purchasing, along with all the grain and hay you cannot grow, lots of fertilizer, pesticides, parasiticides, and antibiotics. It also entails finding a place to dump vast amounts of animal excrement. None of these problems is easy to solve, and each presents its own cascade of effects. Industrial agriculture does not solve for pattern. A different model is followed at Polyface Farm, and it involves some ingenious technology.

Polyface Farm began in 1961 when William and Lucille Salatin went looking for farmland on the western edge of Virginia's Shenandoah Valley. They ended up buying "the most worn-out, eroded, abused farm in the area."[21] Tenant farmers had exploited the land for a century and a half, leaving it in abysmal shape. After 150 years of intensive cultivation and plowing, most of the once fertile acreage had been eroded. In many places, there was no topsoil whatsoever; it had been washed away down the numerous gullies—some 14 feet deep—that scarred the property. The Salatins aimed to heal the land, make a good living, and do what they most enjoyed doing: living, working, and raising a family on a farm.

To accomplish these goals, the Salatins practiced polyculture or permaculture (permanent agriculture). Polyculture is an ecologically grounded method of farming oriented toward the imitation of natural systems and cycles. To begin the healing of the land, the Salatins planted trees and allowed the north-facing hills to return to forest. On sunnier land, they developed pasture. Then they set to figuring out how to keep their animals well fed and healthy. Well-fertilized, parasite-free land that would grow plenty of nutritious grass was needed so their animals could forage. For that, the Salatins needed technology.

The technology was remarkably simple: electric fencing and portable chicken coops. The cows on Polyface farm eat only fresh forage, so there is no need to buy expensive grain. To ensure that the cattle do not denude the land through overgrazing, they are herded into pasture paddocks surrounded by an electric fence. Each day, the fencing and the cows are moved. After the cattle have left a piece of pasture neatly trimmed and littered with cowpies, the Salatins haul a portable chicken coop onto it. The chickens, which could not feed in tall grass, feast on the fresh sprouts and easily accessible insects on the cattle-mowed pasture. In the wild, many birds follow herbivores to dine on the insects they attract and expose. Polyface chickens are helped to achieve what their wild cousins do naturally.

The Salatins wait three days before bringing the portable coop to a paddock once the cows have left. Fly larvae, which are found in abundance in cowpies, typically hatch in 4 days. So the chickens arrive just in time to pick out all the juiciest, fattest larvae from the manure. By dining on these grubs, the birds are getting fat while effectively sanitizing the pasture and ridding the farm of pesky, disease-carrying flies. As a result, the cattle at Polyface do not require parasiticides or worming treatments. The chickens are moved to another paddock when their scratch dance on the abundant cowpies has achieved its purpose. Beyond spreading out the manure and sanitizing the land, the chickens contribute their own, nitrogen-rich droppings.

Michael Pollan, who visited Polyface Farm in conducting research for his best-selling book, *The Omnivore's Dilemma*, describes the paddocks vacated by the defecating birds as resembling a "Jackson Pollock painting."[22] To be sure, there is much cooperative artistry at Polyface. When the cows return to a paddock after the chickens leave, their prior grazing and fertilizing and the birds' own nitrogen enhancement have left the grass better rooted, thicker, and more nutritious. Similar symbiotic relationships have been established between the farm's rabbits, earthworms, and chickens and, in winter, between the pigs, cows, and the compost. Equally simple and inexpensive technological innovations and procedures have been employed to establish these relationships. In each case, the imitation of natural cycles has produced or strengthened a symbiotic relationship between species that exhibits both ecological and aesthetic balance.

It would be easy enough to upset this balance. One might let the external market rather than the land's own natural cycles dictate what and how much to grow. Joel Salatin, the son of William and Lucille, remarks, "I could sell a whole lot more chickens and eggs than I do. They're my most profitable items, and the market is telling me to produce more of them. Operating under the industrial paradigm, I could boost production however much I wanted—just buy more chicks and more feed, crank up the machine. But in a biological system you can never do just one thing, and I couldn't add many more chickens without messing up something else."[23] Adding more chickens would necessitate putting more chickens for a longer time on each paddock. Having consumed the grubs, the birds would begin to denude the land of the rest of its organic material, packing down the earth and leaving it scorched by their highly nitrogenous droppings. With more nitrogen deposited than the increasingly bare, hardscrabble land could absorb, the first rain would send highly polluted runoff into local watersheds. Raising more chickens would not just increase the sales of birds and eggs (and only temporarily at that); it would pollute the water, erode the land, detract from the nutritional quality of the grasses, and degrade the sustainability of the farm.

The Salatins insist that they are not just in the chicken business, the egg business, or the beef, pork, or rabbit business. They see themselves as being in the grass business, the forest business, the compost business, and the earthworm business. Indeed, from the point of view of each of these organisms and the organic processes occurring at Polyface, the farm provides an *oikos* that facilitates a rich and sustainable network of life.

Industrial models of farming follow a linear model of action, executing specific efforts to achieve specific results. At Polyface, the relationship between the plants, animals, and biological processes forms a "loop rather than a line, and that makes it hard to know where to start, or how to distinguish between causes and effects."[24] At industrial farms, chicken droppings are an unwanted, toxic side effect of the explicit effort to fatten up hens for the market. Disposing of the excrement is an expensive and ecologically challenging task. At Polyface, there are no side effects: everything that happens contributes to a self-reinforcing cycle. Chicken manure is not the undesirable by-product of raising birds; it is the cause of the lush pastureland that helps grow beef. Loosing cows on pastures, in turn, is not the isolated effort to fatten herds. It is the preparation of newly mown, grub-laden fields for hungry chickens. At Polyface, farming does not impose linear, technologically driven processes aimed at singular results on a multidimensional network of life. Rather, it reinforces cyclical, self-sustaining interactions. Whenever they forage, eat, or defecate, Polyface chickens, cows, pigs, rabbits, and earthworms contribute to and strengthen a web of relationships.

Polyface Farm annually produces 40,000 pounds of beef, 30,000 pounds of pork, 25,000 dozen eggs, 20,000 broilers, 1,000 turkeys, and 1,000 rabbits for local consumption on 100 acres while maintaining the land's fertility and sustaining a family for three generations in what it loves doing. The Salatins could not achieve all this without technology. Electric fencing and a portable chicken coops are neither highly sophisticated inventions nor feats of unsurpassed engineering. They are simply sustainable solutions for the task at hand. They solve for pattern.

Wendell Berry, a farmer himself, observes that "good solutions exist only in proof, and are not to be expected from absentee owners or absentee experts. Problems must be solved in work and in place, with particular knowledge, fidelity, and care, by people who will suffer the consequences of their mistakes. There is no theoretical or ideal *practice*."[25] As the Salatins demonstrate, solving agricultural problems is not a theoretical enterprise that can be decided from afar, once and for all. It is an exercise in practical wisdom, carried out in the context of local circumstances, needs, opportunities, and knowledge.

Appropriate Technology

The effort to develop and use artifacts and processes sustainably, in a manner that solves for pattern, is often given the monikers of *alternative*, *soft*, or *appropriate* technology. Appropriate technology allows us to better meet our needs through invention and ingenuity while exercising responsible citizenship within the community of life. The visionary economist E. F. Schumacher (1911–1977), whose work I address more extensively in the next chapter, suggested that the best technology was "intermediate," standing between indigenous, customary methods or artifacts developed from local practices and cutting-edge but often out-of-scale innovations. Intermediate technology, for Schumacher, must be inexpensive and accessible, suitable for small-scale application, and compatible with the human need for creativity.[26]

Most examples of appropriate technology are simple feats of engineering that do not require expensive or sophisticated components, are frugal in their use of energy, and have a low impact on the natural environment. In turn, they are attentive to the "reversibility principle," which is to say, they are considered "safe to fail."[27] The reversibility principle holds that when we consider developing or deploying a new technology, only those products or processes should be endorsed whose negative impacts will cease and can be reversed once the technology is withdrawn. Technology that abides by the reversibility principle is safe to fail. That is, if it does not work as planned or has intolerable side effects, it can be terminated. And once terminated, there will be no major, ongoing impacts. Appropriate technology can be innovative, but it does not release a genie from a bottle. It can be experimental without being irresponsible.

Appropriate technology is appropriate to a local habitat or *oikos,* to a specific community. It cannot be taken out of context and retain its value. The community in question, however, does not have to be defined by geographical boundaries. Indeed, it does not have to be defined by the physical interaction of its members at all. Consider the virtual community of people who use the Internet.

The Internet is probably the most important technological innovation in the field of communication since the invention of the movable-type printing press by Johannes Gutenberg in 1436. It has created a plethora of emerging, expanding, and dissolving virtual communities. These webs of relationships produce and share information, ideas, sounds, and images by way of the written word, graphics, photographs, videos, the spoken word, music, and interactive multimedia. This global grid of computer-based networks, a so-called network of networks, allows the virtual interaction of more than 20 percent of the world's population through a planet-wide system of copper wires, fiber-optic cables, and wireless connections. Its infrastructure of computer hardware and software has allowed the creation of a World Wide Web, the vast and ever-expanding collection of documents, images, and recordings connected through hyperlinks and URL (universal resource locator) addresses.

Many users of the Internet and World Wide Web do not view themselves as members of a cyberspace community. They are simply individuals using a modern communication tool. But many self-conscious communities have formed through the Internet. Some of these virtual communities create and maintain collaborative Web sites, called wikis. Wikis may be one of the best examples of appropriate technology that escapes geographical boundaries.

A wiki is a Web site or collection of linked Web sites that anyone can contribute to or modify. There are countless wikis on the Internet today. Appropedia.org, for example, is an "appropriate technology wiki," designed to facilitate the sharing of knowledge and collaboration in pursuit of sustainability. The most popular wiki (and the second largest in the world, after the Chinese language wiki.cn) is Wikipedia.

Launched in 2001, Wikipedia is the largest and most frequently used general reference work on the Internet and the most comprehensive encyclopedia ever assembled. It has been described as "the modern equivalent of the Alexandrian library whose purpose was to make the sum of the world's knowledge available to everyone in one location."[28] Indeed, Wikipedia's goal, its founder Jimmy Wales states, is to provide "every single person on the planet . . . free access to the sum of all human knowledge."[29] To that end, Wikipedia currently consists of over 10 million articles, and the number grows daily. These articles are compiled and edited by more than 75,000 contributors in 253 languages (with English the predominant tongue). Its volunteer contributors write, edit, and update its articles at all hours of the day and night, 7 days a week, 52 weeks a year. A paid staff of fewer than 35 employees of the Wikimedia Foundation, working regular hours, takes care of administrative and legal issues. Wikipedia is currently the fifth most-read Web site in the world, attracting well over 600 million users annually. In terms of popularity, this puts Wikipedia in the ranks of search engines, video sharing sites, and social network sites such as Yahoo! Google, YouTube, MySpace and Facebook.

Wikipedia is well designed for a virtual community in contemporary cyberspace—a vast, ever-expanding, and increasingly interconnected realm of aural and visual communication. It is free, open source, and collaboratively produced. This is not to say that Wikipedia does not have limitations, cause problems, or generate unintended consequences. Its older, well-established articles tend to be accurate, balanced, well edited, and of a quality approaching or even surpassing those appearing in standard hard-copy encyclopedias.[30] But newer Wikipedia entries often betray significant oversights and lack proper citations. They may contain falsehoods and misinformation. These problems, including deliberately misleading claims and vandalism, may remain on a site for months before being detected, deleted, or corrected.

A well-known side effect of Wikipedia (well known, at least, to college professors) is the product of its most salient virtues: its accessibility and comprehensiveness. Owing to its ease of access and incredible scope, students tasked with research projects who would never dream of relying on a hard-copy encyclopedia as their primary source will produce term papers largely, if not solely, based on Wikipedia's online entries. Little other research is conducted to verify or deepen their investigations. Students looking for sources of data, historical information, or lines of argument may find Wikipedia too good to be true—and they are right. Anything that can be used can be misused, and Wikipedia is no exception. Indeed, the Internet as a whole, by providing a cornucopia of information and entertainment to be surfed and consumed in bite-size chunks, may contribute to undermining our capacity for sustained concentration, intensive reading, and contemplative thinking.[31] Still, Wikipedia fits the bill of appropriate technology. There are four major reasons.

First, using Wikipedia does not require overly expensive or sophisticated instruments. In an age of relatively cheap and publicly accessible computers, virtually anyone can exploit its resources. To be sure, impoverished people who have no means of accessing the Internet are excluded. But Wikipedia does not contribute to their impoverishment and may aid in abating it. In any case, Wikipedia makes information easier, not more difficult, for impoverished people to access.

Second, Wikipedia is, comparatively speaking, energy frugal, and has a low impact on the physical environment. Compared to the constant printing and distributing of updated hard-copy encyclopedias, Wikipedia is ecologically friendly.

Third, Wikipedia is safe to fail. If it ceased to be useful, or proved counterproductive, the site could be removed or would simply die of neglect. To be sure, if Wikipedia had to be abandoned, there would be a tremendous loss of easily accessible knowledge. But after the tragic loss to flames of the Royal Library of Alexandria, life went on much as before, if a little poorer.

The death of Wikipedia would be no less tragic, but no more devastating. In any case, it would likely be terminated only if a better means of sharing knowledge arose.

Fourth, Wikipedia benefits a community through the increased interaction of its members without undermining the social, economic, or ecological networks that allow this community to be sustained. Wikipedia is a solution to a particular problem: the need for free, easily accessible, and ever-increasing amounts of information, regularly updated in a quickly changing world. It solves this particular (and significant) problem in a manner that builds a community of volunteer contributors and beneficiaries. Its solution is grounded in an appreciation of, and helps to strengthen, a network of relationships.

The Question of Control

A distinguishing feature of appropriate technology is that it does not create or heighten the dependencies of the members of a community on technological systems that they do not, or cannot hope to, control. The Salatins did not generate the electricity required to run their portable fences. They were already hooked up to the grid when they adopted this technology. Employing a portable electrical fence did not heighten their dependency on a power system that they did not control. And the task of generating their own electricity, though undoubtedly inconvenient and uneconomical, would be feasible. Likewise, users of Wikipedia do not control the Internet or the World Wide Web. No one does. But users of Wikipedia do not automatically heighten their dependency on cyberspace. And unlike readers of books and other sources of information, Wikipedia users are all potential, if not actual, contributors and editors, and thus they collectively maintain a significant level of control over the site's content and development. Inappropriate technology, in contrast, makes people more dependent on resources or systems they do not currently control and cannot reasonably expect to gain control of. Consider a few examples.

Although breast milk is widely recognized as the healthiest food for babies, infant formulas, employed properly, have their uses and benefits. In the 1960s, as birth rates rapidly rose in developing countries while declining in industrialized nations, corporations that manufactured infant formulas increased their marketing campaigns overseas. Potential

customers were led to believe that using the product would produce health-
ier, happier babies. At times, free packets of powdered formula were dis-
pensed in maternity clinics, effectively encouraging the immediate weaning
of babies. The results of the marketing were impressive, with bottle feeding
increasing to 50 percent of all newborns in some countries.

In industrialized nations, the use of infant formulas, while not nearly
as beneficial as breastfeeding, did not produce a crisis. The consequences
in developing countries, however, were catastrophic. Impoverished parents
often watered down the formula to make it last longer, depriving their
infants of the nourishment they needed for health and growth. Even more
problematic, many parents lacked clean water for use with the powdered
formulas or the ability to sterilize water and used bottles. Consequently,
bacterial infections and diarrheal malnutrition became a leading cause of
infant mortality in many developing nations. In response to these problems, protest groups and lawsuits in the 1970s halted many of the advertising campaigns, and some nations banned the promotional practices.[32]

Apart from its nutritional inferiority, the chief problem with infant
formula is that it makes many parents dependent on resources they do not
control: the availability of clean water, means of sterilization, and cash to
purchase adequate amounts of the product. Given the option of breastfeeding, which is available to virtually all mothers, infant formula is not an
appropriate technology, particularly for impoverished people.

The prospect of geoengineering the atmosphere as a solution to the dire
problem of global warming presents a challenging confrontation with the
meaning and use of appropriate technology, and it cuts directly to the issue
of control.

Advocates of geoengineering include one of the greatest living scientists,
a scholar with impeccable academic credentials and the highest environmental standards. Paul J. Crutzen, an atmospheric chemist, won the Nobel
Prize in chemistry in 1995 for his pioneering efforts to identify the causes
and dangers of stratospheric ozone depletion. He remains one of the most
cited and respected scientists in geology. *Time* magazine honored him in
2007 as one of the foremost "heroes of the environment," and James
Hansen, director of the NASA Goddard Institute for Space Studies, identified Crutzen as the "chief scientific caretaker of life on the planet."[33]
Crutzen (along with Eugene Stoermer), proposed the term *anthropocene* to
describe the most recent geological epoch, beginning with the industrial

age in the latter part of the eighteenth century. The anthropocene describes a period of time where the global effects of human activities are clearly noticeable within many different earth systems. One of the primary pieces of evidence for the onset of the anthropocene is global climate change. Crutzen predicts that climate change will have catastrophic consequences for many species on the planet, including human beings. Deeming current efforts to reduce the production of greenhouse gases wholly inadequate to the scale of the problem, Crutzen believes we must craft an "escape route" before planetary warming careens out of control. He has proposed that we fight climate change by altering the chemical makeup of the stratosphere.

In 1991, the volcanic eruption of Mount Pinatubo in the Philippines, spewed 20 million tons of sulfur dioxide into the atmosphere. Scientists noticed that this phenomenon caused an appreciable cooling of the planet, up to 0.5 degrees centigrade the following year, as the chemical oxidized to form sulfate dust particles that acted like tiny mirrors in the stratosphere reflecting sunlight back into space.[34] Crutzen is one of a number of scientists who think we could achieve significant cooling of the planet by reproducing this stratosopheric albedo effect on a regular basis. The idea is to introduce large amounts of sulfur dioxide into the upper atmosphere by employing either a fleet of high-altitude balloons fed by hoses or a battery of heavy artillery guns firing shells. Over 5 million tons of sulfur dioxide a year, about a quarter of what Pinatubo emitted, would be necessary to offset the effect of rising greenhouse gases. Estimated costs of stabilizing planetary temperatures for two years employing these methods range between \$25 and \$50 billion.[35] That is no small sum, but it is dwarfed by the estimated costs of the damage that climate change will likely produce.

Crutzen insists that quickly lowering carbon dioxide emissions is the best response to global warming. In this way, we could avoid the need for his "stratospheric albedo enhancement scheme." However, Crutzen also believes that the world's leaders will not muster the political will to reduce greenhouse gas emissions sufficiently. To believe otherwise is, in his estimation, a "pious wish." Realistically, there are no alternatives to a "large-scale climate modification . . . experiment."[36] As this example demonstrates, people with the highest intelligence, impeccable scientific credentials, and the most sincere environmental convictions may embrace large-scale, sophisticated technology as the only answer to the most pressing problems. Effectively they reject solving for pattern as too slow or too demanding of resources or capacities (e.g., political will) in short supply.

Reviewing the proposal for atmospheric geoengineering, Indian climatologist Rajendra Pachauri, chairman of the 2,000-strong U.N. network of scientists grappling with climate change, warns that "if human beings take it upon themselves to carry out something as massive and drastic as this, we need to be absolutely sure there are no side effects."[37] Backers of the geoengineering effort claim that it presents minimal risks. Crutzen maintains that if untoward consequences do occur, the procedure could be quickly halted, with atmospheric conditions returning to normal within a few years. Ironically one of the potential side effects of spraying our atmosphere with sulfur dioxide would be the increased destruction of stratospheric ozone, precisely the problem that Crutzen had so brilliantly worked to identify and combat in the 1990s. Other potential side effects are the topic of much speculation and study.

Atmospheric geoengineering presents a classic case of a technological solution (spewing massive amounts of sulfur aerosols into the atmosphere) designed to fix a specific problem (global warming) that is itself a product of another technological solution (the extraction and burning of fossil fuels to create energy and heat for a growing human population). The geoengineering solution, in turn, will produce other problems—ozone depletion, acid rain, and, quite likely, unintended consequences yet to be identified—that future scientists and engineers will have to remedy. Indeed, its most deadly side effect might be the depletion of political will and hindering of efforts to reduce carbon dioxide emissions. As the threat of global warming is deferred through atmospheric geoengineering, political leaders may find it harder to convince their constituents to wean themselves from fossil-fuel-based economies and lifestyles.

Geoengineering the atmosphere does nothing to actually reverse, or even slow, the production of greenhouse gases. It simply compensates temporarily for their heat-trapping tendencies. Once geoengineering efforts stop, all of the built-up greenhouse gases will exert their full effect. In this respect, spraying sulfur aerosols into the sky to combat climate change has been likened to fighting a bad case of obesity with a corset and a diet of doughnuts.[38] Things may seem back in control for the time being, but we do not want to imagine the results when the corset breaks. And there may be little opportunity, or time, for reversing course when the seams start to tear.

Geoengineering the atmosphere also does nothing to address the acidi-fication of the oceans caused by increased amounts of atmospheric carbon dioxide. The oceans absorb up to 40 percent of the carbon dioxide that we emit into the atmosphere. This process tamps down the greenhouse effect, but at a price. Oceans today are already 30 percent more acidic than they were in preindustrial times, and their acidity is steadily rising. Corals and shelled animals are unable to use the calcium carbonate they need to grow in acidic waters. The rising acidity of our oceans portends dire conse-quences for sea life and the human populations that depend on it.[39]

Geoengineering is a technological quick fix. It temporarily abates one problem (global warming) without addressing its underlying causes (green-house gas emissions) or their concomitant negative effects (acidification of the oceans). And it may reduce the effort and capacity to grapple with underlying causes. Importantly, geoengineering the atmosphere cannot be locally controlled. If it is carried out, the entire population of the planet will be subject to its effects and side effects. One could argue that geoen-gineering is necessary to avoid, or defer, the calamity of rising global temperatures. But it is not appropriate technology.

So-called terminator seeds, and potentially much other biotechnology, provide a final example of inappropriate technology. Terminator seeds were developed by the Monsanto Corporation, a multinational business that produces herbicides as well as hybridized and genetically modified seeds for agriculture. In 1998, Monsanto secured a patent on seeds devel-oped with genetic use restriction technology (GURT), colloquially known as terminator technology. GURT seeds grow into plants that produce sterile seeds: the seeds from GURT plants can be eaten if harvested but will not grow if planted. A billion farmers worldwide, most of whom live in devel-oping countries, save a portion of the seeds from their crops each year for planting. Sterile-seed technology would end this age-old cycle, protecting the property rights of the biotechnology company and effectively forcing farmers to buy new seeds every season.

Critics charge that terminator seeds would transform formerly indepen-dent farmers into "bioserfs." Defenders of the technology insist that farmers are not forced to purchase the product and are free to use their own seeds. The problem is that large seed companies might create virtual monopolies, such that (poor) farmers find themselves with a shrinking pool of nongenetically modified seeds. After a storm of protest, Monsanto agreed not to commercialize its terminator technology. But it has continued to buy corporations that have or are developing GURT patents. Critics worry that Monsanto and other multinational corporations that are in the GURT business are simply waiting for the furor to die down before reintroducing their products into the marketplace.[40]

GURTS are not the only genetically modified organisms (GMOs) that have been developed and patented. Monsanto and many other companies market a long list of genetically modified plants. Arguably the use of any of these GURTs or GMOs significantly heightens the dependence of members of a community (in this case, farmers) on a technological system they do not control. In theory, refusing to buy or use GURTs and GMOs is always an option. In practice, the situation for farmers is more complex and challenging.

It certainly proved more complex and challenging for Percy Schmeiser, a Canadian farmer whom Monsanto sued. Born in 1931, Schmeiser grew and bred canola plants in the prairie province of Saskatchewan for half a century. Then in 1998, Roundup Ready canola plants were discovered in Schmeiser's fields, and Monsanto sued for patent infringement. Roundup Ready crop seeds, one of Monsanto's signature products, have been genetically modified to survive large applications of Roundup, a herbicide that Monsanto also developed. These seeds contain "in-plant tolerance" to the herbicide. Farmers can use Roundup generously and frequently to kill weeds without deterring the growth of their genetically modified crops.

Schmeiser claimed that his fields were cross-pollinated from nearby crops of Roundup Ready canola and fought Monsanto's charge in court. The case worked its way through the legal system for over six years. After mortgaging his farm to pay legal fees, Schmeiser saw his case go before the highest court in the land. In 2004, the Canadian Supreme Court ruled that a farmer does not own patented, genetically modified seeds grown on his property even if they have been blown on his land from, or were pollinated by, nearby fields. Schmeiser could not continue to use seeds from his canola plants if they contained the Roundup Ready gene without paying Monsanto its established fees. Left with seeds he could not harvest and replant, Schmeiser eventually stopped growing canola altogether, shifting to wheat and other crops.[41]

Schmeiser's case illustrates the challenges faced by farmers in a world of genetically modified organisms. If they opt to grow GMO plants, they become dependent on technologies they cannot control. And if they do not buy GMO seeds, they may find themselves constrained to adapt to a world significantly altered—some might say infected—by GMO technology. Many farmers, perceiving the threat of lengthy and costly lawsuits, take the path of least resistance and adopt GMO agriculture. They cannot afford to do otherwise. To add insult to injury, recent research suggests that genetically engineered seeds do not actually increase yields that farmers receive from their food and feed crops.[42] And the extensive use of herbicides on GMO crops contributes to the development of superweeds.

Farmers are not the only people facing such problems. Soon parents may confront similar conundrums. While moratoria exist on many forms of human genetic enhancement, the genie may already be out of this bottle. The science of identifying particular human genes linked to specific human traits or capacities is far from well established. Still, we might realistically envision a world, perhaps in the not-too-distant future, where human embryos are genetically modified to add, increase, or strengthen particular physical and mental attributes and abilities. Facing a world of genetically altered tots who have greater memory, stronger analytical abilities, and quicker wits, will soon-to-be parents leave the developmental potential of their children to the whims of nature and the slow road of education? Will couples really feel at liberty to bear a "natural" child if he or she proves genetically disadvantaged at school and, subsequently, in business and professional circles? Even those who shudder at the thought of this brave new world might effectively be forced to embrace genetic modification to prevent their children becoming second-class citizens in a world of techno-tots.

The social and political ramifications of human genetic modification are dire. Progeny of those who could afford the latest technological tweak will become the best and the brightest, while the rest of the population will be left behind. A caste system would develop, and harsh inequities might arise even within families. With engineering advancing at a quickening pace, parents of a genetically enhanced child might anguish over whether his or her future siblings should be the same model or benefit from new and improved technology. In a socially and economically competitive world, Bill McKibben muses, parents who do not pony up for genetic enhancements for their offspring might well feel culpable of "child abuse." Yet the prospect looms that accelerating technological improvements will leave parents who enter the "biological arms race" with an outdated child, the equivalent of "a nearly useless copy of Windows 95."[43]

Such moral quandaries are mind-boggling, and they do not even broach the issue of malfunctions. While genetic enhancement technology is being perfected, one will undoubtedly witness the birth of many "Frankenstein" children condemned to horrible fates. But here, as with most other technological projects, the most crucial question to ask is not, "Will it fail?" but rather, "What else will it do if it succeeds?" Success in our technological efforts, not malfunction, may produce the most frightening consequences and conundrums. At the end of the day, our efforts to control technology may well result in technology controlling us. It may determine not just what we do but, quite literally, who we are.

Letting Things Be

From a technological point of view, the earth appears as a storehouse of resources to exploit and control. The only question that the technologist confronts, as a technologist, is how to devise the most efficient means to achieve his ends. And any given end—including the destruction of life or its genetic re-creation—can be pursued efficiently. The technological frame of mind encourages us to leave ethical, political, philosophical, and spiritual concerns at the door of the laboratory or factory as we enter. Without practical wisdom to guide choices, technological ingenuity inevitably shakes open Pandora's box.

It is naive to believe that, having opened Pandora's box, we can design failsafe systems to prevent pernicious outcomes. Notwithstanding his strong advocacy of nuclear power, Edward Teller, a participant in the Manhattan Project and so-called father of the hydrogen bomb, reputedly said, "There's no system foolproof enough to defeat a sufficiently great fool." We cannot predict what might go wrong with a technological system, but we can predict that humans will continue to make foolish mistakes—and some of them will be doozies. Given this reality, we would be well advised to engage in safe-to-fail experimentation rather than hanker after failsafe systems. Appropriate technology is a safe-to-fail means of satisfying human needs, including the need for adaptation in a changing world.

The challenge we face today is not simply to better control technology, though this is a crucial task and one that is hardly assured success. But if we focus solely on controlling the processes, tools, and machines that we produce through technological means, at best we will succeed in becoming better technologists. Success in this endeavor would likely have the effect of reinforcing the hegemony of a technological worldview. In this respect, the most important characteristic of solving for pattern is that it educates us differently. It prompts us to focus on patterns—networks of interdependent relations—as much as, if not more than, the specific technological issue at hand. Solving for pattern underlines our membership in social and ecological communities; it does not simply make us more technically astute at manipulating our world. Ultimately, if developing and using appropriate technology does not stimulate a greater level of ecosophic awareness, it is not appropriate enough.

Heidegger writes, "So long as we represent technology as an instrument, we remain held fast in the will to master it."[44] His point is that our lives must not be wholly circumscribed by efforts to control the world. We have to forgo the Midas touch that turns everything into something of use and break free of the Midas mind that sees everything as a potential resource awaiting our touch.

Technology is an important, even crucial part of human existence. It now threatens to become the whole. By developing a greater understanding of the interdependencies that characterize the world, we can cultivate the practical wisdom required to make use of appropriate technology. In this manner, we might skillfully employ the part (a technological artifact or process) in the context and for the benefit of the whole (the *oikos* or community of life). By developing appropriate technology that solves for pattern, we escape the futile and perilous quest to do just one thing. Heidegger wrote:

We can use technical devices as they ought to be used, and also let them alone as something which does not affect our inner and real core. We can affirm the unavoidable use of technical devices, and also deny them the right to dominate us, and so to warp, confuse, and lay waste our nature. But will not saying both yes and no this way to technical devices make our relation to technology ambivalent and insecure? On the contrary! Our relation to technology will become wonderfully simple and relaxed. We let technical devices enter our daily life, and at the same time leave them outside, that is, let them alone, as things which are nothing absolute but remain dependent upon something higher. I would call this comportment toward technology which expresses "yes" and at the same time "no," by an old word, *releasement toward things*.[45]

The "old word" Heidegger refers to, in its original German, is *Gelassenheit*. He borrowed the term from Meister Eckhardt, a thirteenth-century German mystic and theologian. *Gelassenheit* literally means "letting be."

Releasement entails letting things be in their multiple and diverse relationships rather than appropriating them as just one thing: resources for our use. To let the things of this world be in their multiple and diverse relationships is to acknowledge that we do not fully comprehend and cannot ever control the intricate interdependencies they represent. In turn, releasement toward things entails a conscious and deliberate restraint. With the end of slavery, humanity came to understand that not everything that could be bought and sold should be bought and sold. It is possible that a new relationship to technology will convince us that not everything that can be put to use should be put to use.

Nowhere is this restraint more pressing than in our current relationship to fossil fuels. The exploitation of fossil fuels ushered in an era of unprecedented technological development and wealth. Employing an apparently endless and highly efficient energy source buried beneath the earth's crust, our species was able to turn the planet into its workshop and extort from it ever more resources. Virtually everything on and under the earth could be put to productive use through the magic of these highly combustible materials. Indeed, fossil fuels might be seen as a gift from Prometheus. Their fires, burning steadily now and at an accelerating rate for two centuries, have given us dominion over the earth. But our planetary workshop is now getting uncomfortably hot.

So we face a choice. Do we burn every last ounce of fossil fuel that we can get our hands on? Or do we let it be? And would our letting it be— letting the remaining fossil fuels remain fossils, deep within the earth— mark a new relationship toward ourselves and the planet? Perhaps for the first time in human history, we would step away from the full exploitation of an available and valuable resource. To do so, we would have to cease seeing the oil and coal buried under the earth's crust as standing reserve. This may prove to be the first collective step we take as a species toward letting other things be, no longer appropriating everything as a resource for our benefit. It is possible that the pressing need to stabilize climate will present humanity with a unique opportunity to release itself from the grip of a technological way of thinking and acting. The opportunity, ironically, arises as the civilization we built with Prometheus's gift is most threatened by it.

Conclusion

Sixty-five million years ago, a good-sized asteroid, 6 miles in diameter, was flying about the galaxy minding its own business when a planet got in its way. The impact caused an explosion equivalent to the blast of 100 million megatons of TNT. It instantly vaporized the asteroid and a sizable chunk of the planet's surface, creating a crater over 100 miles wide. The adjacent seas began to boil, and firestorms raged across the globe as debris cast up into the heavens reentered the atmosphere as incandescent rock. Once the conflagrations died down, the skies darkened with soot to the point of permanent night that lasted for months, if not years. Three-quarters of all the species on the unfortunate planet were exterminated by its chance encounter with a flying rock.

The planet in question, of course, was Earth. The asteroid's impact off the Yucatan peninsula of Mexico some 65 million years ago, many scientists argue, brought about the extinction of the dinosaurs, creatures that had dominated life on earth for some 160 million years.[46] This was neither the first asteroid to hit our planet nor the biggest. Most certainly, it will not be the last. Indeed, scientists estimate that our chances of being struck by an asteroid of significant size within the next century are about 2 percent. Expand the time horizon, and our chances increase accordingly. Eventually a catastrophic impact is inevitable.

If there is any reasonable hope of avoiding such catastrophe, it will come from technology. Concerned scientists and engineers are already hard at work. The B612 Foundation (which is named after the asteroid upon which Saint-Exupéry's Little Prince lived) has as its mandate the alteration of an asteroid's orbit by 2015.[47] Saving our planet from an asteroid impact may not be the most pressing need in an era of rapid climate change and massive species extinction, but it is by no means a worthless enterprise. Such heroic efforts underline the fact that technology—which has caused so much destruction on our planet—has also done much to safeguard life, and will be asked to do a great deal more. The sustainability of civilization, it is fair to say, depends on the kinds of technology we choose to develop.

"The rise of the human neocortex," novelist Arthur Koestler (1905–1983) observed, "is the only example of evolution providing a species with an organ which it does not know how to use."[48] Our technological efforts demonstrate this fact with terrifying force. We flirt with power of stupefying magnitude. Nuclear fission, genetic manipulation, geoengineering, and nanotechnology are salient examples. The neocortex makes us technologically clever, and this cleverness often gets us into trouble.

Certainly we are in some tough predicaments today. But we cannot abandon technology any more than we can undo our evolutionary development. Our intelligence and technology will play important roles in finding ways out of our predicaments. We can and must develop technology with greater intelligence or, perhaps better said, with greater wisdom. Knowledge makes us clever and allows us to do many things. Wisdom places these abilities in the context of their consequences and our limitations. We are too clever today to survive without becoming wise.

To an unprecedented degree, the human species modifies the world in which it lives. Increasingly, we attempt to create this world anew. When we employ technology, we design a course of action to achieve a single, preconceived purpose. Technology helps us get much accomplished in the most efficient manner. But focusing narrowly on efficiency and engineering while neglecting broader relationships has its costs. If we were squirrels, a technologist might come up with a drug or computer chip to enhance our memories so we could find all our hidden acorns. We would quickly enjoy greater nourishment and more leisure time. And in a few generations, absent the regeneration of oak forests, we would perish—or face the task of engineering new sources of shelter and food. Heightened efficiency and innovation does not shield us from the first law of ecology. Indeed, it often blinds us to this law and increases the penalty for ignoring it. Solving for pattern means no longer using technology in a blinkered effort to do just one thing. It requires that we employ our technological knowledge and skill to meet a higher standard: sustaining in multiple, synergistic ways the networks of life that sustain us. The question is whether we have the political and social will, the patience, and self-control—in a word, the practical wisdom—to solve for pattern.

Zeus punished Prometheus severely for his thievery. Shackled to a peak in the Caucasus, Prometheus is tortured by a vulture that feasts on his liver. Each night, Prometheus's organ grows back that it might be devoured again by the voracious raptor the following day. Prometheus is bound to a rock and suffers this terrible fate because he could not limit himself, a trait shared by the technological race he created. Perhaps Prometheus has his liver eaten each day because it is the organ of envy. Promethean man is an ambitious and envious creature. He wants nothing less than to become a god, the master of his world. His technological powers lead him to believe that apotheosis is indeed possible. To be sure, humankind's feats of engineering are astonishing. Yet each wondrous achievement leaves it ever more aware of limitations yet to be surpassed. Every overcoming of a boundary further underlines the distance between the Promethean race and full divinity. Thus, every dawn we are awakened from our dream of omnipotence by a ravenous lust for greater mastery and control that eats away at our souls.

We live in a world of furious technological growth. There is always something more to do and new to make. There is no end to creating things smaller and faster, bigger and more powerful. The call of the new and improved is a siren song few can resist. In such a world, supply stimulates demand, as excited wants soon become perceived needs. We might wish to believe that a novel device will wholly satisfy a genuine need. In fact, it stimulates more needs, more wants, and the pursuit of successors. Technology has an insatiable hunger. The danger of looking to technology to solve our problems and sate our desires is that we become habituated, one might even say addicted, to technological quick fixes. And like any addict, we may soon find ourselves incapable of thinking any other thought than that of getting our next fix.

As technology becomes our dominant mode of thought and action, other distinctly human ways of being in the world may atrophy. The linear, purposeful drive of technology—so powerful and so useful—may diminish our awareness and appreciation of the interdependencies of life. Should this come to pass, we will become modern-day Midases. Having adopted Prometheus as our patron god, however, we will find it necessary to extort new, enhanced powers every time our previous interventions go awry. The scope and ingenuity of our ever-developing craft will be great sources of pride and wealth. But such technological power will not ultimately sate our deepest hunger or sustain our lives.

We should not spurn technology. Many problems, including some of those caused by technology, demand technological solutions. Like any other species, our survival and prosperity depend on adaptation. And for the human species, adaptation often requires technology. We have an abundance of problems to solve. But we must solve for pattern, and we must exercise practical wisdom in every case.

The real promise of technology is that its vast scope, fecundity, and power push us to confront the boundaries of the human condition and its deepest meanings. There is perhaps no greater stimulant of ecosophic awareness than the threat that technology may one day extinguish it.

4 Economics

There is no such thing as a free lunch.
—Barry Commoner

Dr. Seuss taught me my first lesson in economics. His fanciful tale, *The Lorax*, published in 1971, the year after the first Earth Day, had a lasting impact. The enterprise of the ambitious Once-ler demonstrated that endless economic growth, or "biggering," leads to the destruction of environments and communities. When the Once-ler first arrives in the land of Grickle-grass, the sight of swaying Truffula trees starts the wheels spinning. Soon the softer-than-silk Truffula tufts are being knit into thneeds. No one needed a thneed before the Once-ler started making them. But supply fed demand, and soon the brightly colored Truffula trees were falling four at a time to the Super-Axe-Hacker. The Lorax, "who speaks for the trees," reprimands the Once-ler as his factories keep biggering. He also speaks for the Brown Bar-ba-loots who eat Truffula fruits, the Swomee-Swans who cannot sing their notes owing to the smog in their throats, and the Humming-Fish whose gills are gummed from the effluent glump. But the Once-ler is deaf to the Lorax's call for restraint. Then, with a sickening smack, the last Truffula tree falls. The factories close, the workers leave town, and a desolate land is all that remains.[1]

The Lorax, powerful in its simplicity, teaches a basic economic truth. Producing thneeds or any other commodity is not doing just one thing. It affects the natural world, and the communities that rely on nature's bounty. The message has echoed across the decades. Indeed, a number of Lorax programs remain active today.[2] They teach children, who will soon be adults, to "speak for the trees" and the communities they sustain. But the conviction that the single-minded pursuit of financial gain is environmentally and socially destructive is not limited to fans of the Lorax. The world teaches this lesson daily.

Economics and Sustainability

Environmental destruction occurs for many reasons, not all of them directly related to making money. As the English political theorist Thomas Hobbes argued, human beings are motivated by more than just wealth. They also pursue physical security, and their efforts to protect themselves lead to arms races and warfare. Training, maintaining, and outfitting armies, navies, and air forces consume vast amounts of national resources.

Lester Brown has calculated that the budget for meeting basic social needs globally (including universal primary education, the end of adult illiteracy, school lunches, basic health care, and family planning) and restoring the earth's ecosystems (including stabilizing water tables, biological diversity, fisheries, rangelands, topsoils, forests, and climate change) would be $190 billion annually. That is a great deal of money. But it is $45 billion less than the world spends each year on military personnel and armaments.[3] Developing, maintaining, and using arsenals and armed forces also have a direct environmental impact. By many measures, the largest single polluter in the United States (and the rest of the world) is the U.S. military. And armed confrontations—battles and wars—even beyond their devastating toll in human suffering and lives, are some of the most environmentally damaging activities that occur on the planet.

Likewise, overpopulation is a major contributor to environmental destruction and a significant obstacle to the pursuit of sustainable livelihoods. Debate ranges widely as to the "carrying capacity" of the planet. Some say that the earth cannot indefinitely maintain more than 2 billion people. Some argue the figure is much more, or less. Clearly the earth cannot support continuous population growth. Most acknowledge that moving from the 7 billion people who currently inhabit the planet to a likely 9 or 10 billion within the next half-century will take a huge toll on the environment.

The pursuit of physical security need not lead to large military forces and war any more than the pursuit of reproductive success always leads to overpopulation. When our basic need for physical security and reproduction do cause large-scale environmental destruction, economics is often involved. Wars are frequently fought to gain or maintain access to resources. Powerful business interests work in tandem with national armed forces—the military-industrial complex that President Eisenhower warned about in his famous 1961 farewell address—to ensure the continued prominence of the military and warfare in the world. Likewise, the uneven distribution of economic resources is a significant cause of population growth. If avarice did not bring some to pursue and accumulate so much wealth, others (and in particular, women) might not be prevented from gaining the education and economic opportunities they need to limit family size. Impoverished couples often have many children as a hedge against destitution in their old age, and uneducated, unemployed women often have very large families as a consequence of their powerlessness.

Beyond military activities and population growth, many other factors undermine sustainability. Not all are primarily economic in nature. Yet most, like military activities and population growth, are indirectly, if not directly, related to the pursuit of wealth and vast inequalities in its distribution. The biodiversity of the planet, for instance, is most threatened by the loss of critical habitat to agriculture, logging, and the construction of roads, buildings, and parking lots. Climate change is the direct product of a global marketplace whose engine is fired by fossil fuels. There is good reason to view environmental destruction and unsustainable practices as primarily the products of humanity's long-time love affair with money.

At the height of its empire, it was said that all roads led to Rome. Today all roads to sustainability pass through the realm of economics. To acknowledge this fact is not to celebrate moneymaking as the core of our being. Economics is only part of life, and demonstrably not the most salient feature of a meaningful life. Without using free markets to lift billions of people out of poverty and stimulate the right sort of technological innovations, however, sustainable societies cannot be achieved. And without redefining our relationships to markets and money, and the things they produce and allow us to buy, sustainable lifestyles cannot be achieved. Since the development of economics as a field of study, which accompanied the rise of capitalism in the late eighteenth century, the natural world has been identified as an inexhaustible stockpile of resources to be plundered for gain, and an ever-draining sink into which wastes can be dumped. Pursuing sustainability requires transforming this worldview.

Economic pursuits have led our species to undermine the life-support systems of a finite world. Most certainly, the restructuring of economics will play a major role in any viable effort to keep the planet habitable and hospitable. Historically—and certainly prehistorically—there was a time when economic activities had little effect on human habitat. The planet was big and business affairs small. Today the situation is different. Economic and ecological life are interdependent: "The world's economy and earth's ecology are now interlocked—'until death do them part.'"[4] Transforming our relationship to markets, money, and the things markets and money provide is the only means to sustaining our planetary home. There is no path to sustainability that does not require fundamental reform of how we do business.

Self-Interest, Unintended Consequences, and the Commons

As a field of study, economics concerns the investigation of the laws or rules (*nomos*) governing the place (*oikos*) where the production and provision of goods occur. For the ancient Greeks, the household was the primary place of production: here agriculture and crafts were pursued. Since the onset of the industrial revolution and the expansion of capitalism in the nineteenth century, much of the production of goods moved out of the domestic habitat and its attached agricultural lands to be relocated in urban factories and workplaces.

Adam Smith (1723–1790), the father of the discipline of economics, observed in *The Wealth of Nations* that the individual's selfish pursuit of gain was the driving force of the economic world and the engine of economic growth. "It is not from the benevolence of the butcher, the brewer, or the baker, that we expect our dinner," Smith remarked, "but from their regard to their own self-interest. We address ourselves, not to their humanity but to their self-love, and never talk to them of our own necessities but of their advantages."[5] Each individual seeking his own economic gain, Smith maintained, "is led by an invisible hand to promote an end which was no part of his intention." That end is the public good. The pursuit of individual gain, by way of relations of interdependence within the marketplace, yields benefits for those with whom the self-seeking individual interacts and for the nation as a whole. For Smith, economic life is chiefly defined by relations of interdependence within a market whose hidden hand produces unintended consequences: the public benefit of national prosperity. The field of economics, Smith suggests, is chiefly defined by unintended consequences—and very beneficial ones at that.

Ecology and economics are often understood to operate at loggerheads. Yet one might say that economics is simply the ecology of the marketplace, while ecology is the economics of the natural world. The central truth of ecology, that you can never do just one thing, is also a core feature of the discipline of economics.

Since the days of Adam Smith, economists have heeded the law of unintended consequences; indeed, it is considered one of the building blocks of the discipline.[6] Contemporary economists continue to champion Smith's key insight. They tend to focus on two types of unintended consequences: positive public gains of the individual's selfish pursuit of wealth and the unintended negative consequences of government intervention in the marketplace. Regulation of finance and industry, free market economists claim, generally hurts businesses by hampering their ability to engage in enterprise. Whenever government regulations threaten profit margins, the pursuit of gain that makes for a vibrant market, with all its public benefits, becomes stifled. Restrictions on what can be bought and sold in the market, and regulations as to how these things may be produced, undermine entrepreneurship and the beneficial effects of self-interested economic pursuit.

Proponents of the free market worry about governmental constraints on how wealth can be generated. They also worry about governmental constraints on how much wealth can be accumulated. Taxes are generally seen as a threat to the unrestricted accumulation of wealth and, as such, a threat to economic ambition and its public benefits. In turn, free market advocates worry about governmental constraints on what can be owned. Capitalism is grounded in the private ownership of resources. Public ownership of land, industries, services, or other assets restricts the domain of private ownership and effectively cuts off segments of the world's resources to profitable exploitation.

Recently the privatization of goods designed to facilitate profit seeking has been extended to new reaches. Now it is possible to own entire species and life forms. The landmark 1980 U.S. Supreme Court case, *Diamond* v. *Chakrabarty*, awarded ownership rights to a corporation for a new strain of bacteria developed to break down hydrocarbons and potentially help clean up oil spills. Since then, the patenting of living organisms and their genetic components has been legal. Through genetic engineering, entire new species can be "invented" and patented. Alternatively, "bioprospectors" can mine the genetic resources of a country's flora and fauna for pharmaceutical compounds and other products. In such cases, with the patenting of specific DNA sequences, indigenous communities may find themselves paying royalties for products based on elements of traditional medicine and nutrition that they have been practicing for centuries.[7]

Advocates of free market capitalism argue that privatizing the world's resources is beneficial. Publicly owned resources, they insist, are particularly vulnerable to abuse. The tragedy of the commons described by Garrett Hardin provides the foundation for their argument.[8] In old England, Hardin observed, people grazed their sheep on public land that was called the commons. Livestock owners wanted to take full advantage of the free forage, so every shepherd would steadily increase the size of his flock. Soon the commons was filled with sheep eating the once lush grass to its roots. At that point, the now eroded commons ceased to provide any forage. All the sheep, and all the shepherds, lost out. In this case, individuals pursuing their self-interest to the greatest extent possible did not lead to long-term private or public benefits.

Economists cite the tragedy of the commons to underline the noxious, unintended consequences of the public ownership of resources. That which is not privately owned, the argument goes, suffers most from the unintended consequences of a lack of caretaking. Everyone tries to benefit from the commons, but nobody invests in its conservation. The assumption is that people will only take care of things they personally possess. This is not a novel insight. Aristotle made the same argument over two millennia ago: "That which is common to the greatest number has the least care bestowed upon it. Everyone thinks chiefly of his own, hardly at all of the common interest."[9] To be sure, public goods are difficult to secure and hard to maintain. At the same time, Hardin's classic tale misrepresents the data. Pasturelands, water sources, forests, and fisheries have been communally managed and well conserved throughout history.

Elinor Ostrom, the 2009 Nobel Prize winner in economics, demonstrated that so-called common pool resources across the globe have been sustained through many generations of local stakeholders by way of cooperative ventures and customary practices.[10] In turn, many of the earth's commons today—Antarctica and the stratospheric ozone-layer, for example—are being protected by way of voluntary conventions. Here, as with local, regional, and national resources that are publicly owned and collectively managed, the commons gains protection owing to an appreciation of interdependence. Unlike Hardin's shortsighted shepherds, many leaders and citizens today realize that the unintended consequence of the unencumbered pursuit of private wealth is ecological disaster.

In 1965, shortly before his untimely death, Adlai Stevenson delivered his last speech as ambassador to the United Nations. He likened the earth to a spaceship to underline the interdependencies of planetary life. The following year, economist and systems scientist Kenneth Boulding wrote an essay, "The Economics of the Coming Spaceship Earth."[11] A few years later, architect and futurist Buckminster Fuller wrote his famous *Operator's Manual for Spaceship Earth*.[12] Boulding and Fuller, like Stevenson, emphasized the intertwined fates and common responsibilities of participants of a closed planetary system. In his seminal speech, Stevenson said, "We travel together, passengers on a little spaceship, dependent upon its vulnerable reserve of air and soil; all committed for our safety to its security and peace; preserved from annihilation by the care, the work, and I will say, the love we give our fragile craft."[13] On Spaceship Earth, ecological interdependence sows a common fate and presents a common task for humankind.

The problem, Stevenson observed, was that economic relationships aboard Spaceship Earth belie our common fate and common tasks. Economic interdependence, if adequately addressed, might allow the spacecraft to flourish. If it is ignored, unsustainable economic relations will spell our doom. Echoing Abraham Lincoln's famous "House Divided" speech delivered a century earlier to a nation equally in peril, Stevenson globalized Lincoln's assessment that a "government cannot endure, permanently half slave and half free." In the context of a world approaching the brink of ecological collapse, Stevenson insisted that "we cannot maintain [the spaceship] half fortunate, half miserable, half confident, half despairing, half slave to the ancient enemies of mankind and half free in a liberation of resources undreamed of until this day. No craft, no crew, can travel safely with such vast contradictions. On their resolution depends the security of us all."[14] The passengers on Spaceship Earth, Stevenson was saying, do not equally enjoy the same economic privileges or suffer the same economic deprivations. Yet their ecological fates, and the fate of the craft, are inextricably tied together. The unintended consequence of stark inequalities in wealth and resources—largely though not solely the product of the unencumbered pursuit of wealth—will be the destruction of a planetary habitat.

Great wealth and dire poverty are equally disastrous from an environmental perspective. The former is associated with overconsumption and excessive waste. The latter produces lives of desperation that lead to overpopulation and the despoliation of land. Across the globe, more than 22,000 children die each day owing to extreme poverty; over 1 billion people do not have enough water for basic needs; and many more than half the people in the world live on less than $3 a day.[15] The anthropologist and conservationist Richard Leakey observed that "to care about the environment requires at least one square meal a day."[16] Leakey's point is that the practices of environmental stewardship develop only once basic human needs are met. As important, they diminish again when great wealth insulates us from planetary interdependencies.

Consider the issue of climate change. The United States, the world's wealthiest country, has one of the largest per capita emissions of greenhouse gases. For most of the past half-century, it has been the single largest producer of greenhouse gases in the world. Only the rising economic productivity of China has recently stripped this dubious distinction from the United States. But if great wealth is the most prominent culprit in the crime of climate change, dire poverty proves itself an unwilling accomplice. For example, impoverished farmers in the developing world are often forced to slash and burn rainforests to grow crops. Frequently these crops, such as tobacco (Malawi), palm oil (Indonesia), or soya used for feedstock (Brazil), are primarily grown for export. Apart from destroying critical habitat for endangered species, setting fire to the rainforests—the so-called lungs of the planet—contributes significantly to global warming.

Poverty is not environmentally benign. Wood collected from dwindling forests and eroded savannas fires the cooking pits of a majority of the world's impoverished peoples. The collection of firewood contributes significantly to deforestation and desertification, and hence to global warming. Although electric light bulbs use fifty times less energy per watt than kerosene lamps, many poor people who cannot afford the former are forced to use the latter, contributing even more to greenhouse gas emissions. Well-insulated and well-constructed housing that conserves energy, in turn, is beyond the economic reach of the poorest sectors of most societies. Here, as in the burning of rainforests, the deforestation and desertification caused by wood collecting, and the use of inefficient fuels, impoverished peoples often live and work in ways that exacerbate climate change. And, of course, overpopulation stimulated by poverty worsens all of these trends, while heightening the unsustainable use of natural resources. The vast majority of the world's population growth occurs, and will continue to occur, in developing countries. It follows that the problem of climate change cannot adequately be addressed without confronting the economic pressures that prevent the majority of the world's population from pursuing sustainable livelihoods.

Overconsumption gets a lot of environmental press, and rightly so. But underconsumption is as large a problem. Half of the world's people are not consuming enough to have a decent quality of life. By 2050, according to United Nations' projections, the world's population will increase from our current 7 billion to between 8 and 11 billion. The difference between 8 and 11 billion represents the size of the world's population in 1960, or the population of the entire globe today minus Asia. If the higher figure of 11 billion is reached in the next few decades, underconsumption will surely rise dramatically, as will its accompanying social, economic, and environmental problems. The world's poor want and deserve more, and they are working hard to get it. The tragedy is that if their way of getting more looks anything like how the developed world got more, things are going to get a whole lot worse before they have the slightest chance of getting better.

Vandana Shiva, a physicist and environmental activist, observes that "giving people rights and access to resources so that they can regain their security and generate sustainable livelihoods is the only solution to environmental destruction and the population growth that accompanies it."[17] Climate change brings this relationship to a head. As a result of global warming, glaciers will melt, rivers will dry out in the summer, agricultural production will decrease in key areas, desertification will increase, and low-lying coastlines will flood. Tens, and possibly hundreds, of millions of people will become environmental refugees in the coming decades as a consequence of these climate-induced changes to weather, water supplies, and agricultural production.[18] National borders will not contain these desperate individuals in search of freshwater, food, shelter, security, and livelihoods. Their fate will become an economic reality for neighboring countries and for the world at large. The relationship between our economic interdependence and our ecological interdependence is patent, and ever more precarious.

Economic Interdependence and Globalization

A binge in real estate speculation triggered a credit crisis in the United States in the waning months of 2008. Within weeks, there was a financial meltdown on Wall Street. Notwithstanding unprecedented governmental efforts to manipulate the economy, infusing hundreds of billions of dollars in bailouts, the Wall Street meltdown produced a chain reaction of crises that ricocheted across the planet. Economic turmoil rocked nations worldwide, destroying countless businesses, bringing many national economies to their knees, and setting off a global recession.

Economic models did not predict the financial crisis of 2008 and 2009, which wiped out $50 trillion in global wealth. Indeed, one might argue that these models contributed to the crisis. In its wake, economists are reevaluating their field as a whole.[19] Too large, too complex, and too interdependent for anyone to manage, predict, or fix, the globalized economy is both immensely powerful and inherently unstable. *New York Times* economist Thomas Friedman captured the essence of this precarious planetary web of economic relations with these words: "We're all connected and nobody is in charge."[20] As if to mimic the lessons delivered in recent years in the realm of ecology, the collapse of financial markets demonstrated the complex interdependence of a globalized world.

Economic globalization dates back at least to the intensified trade and finance sparked by European colonization in the 1700s and 1800s. Certainly it was well recognized at the height of the British empire of the early 1900s. With the rise of American power after World War II, economic globalization grew markedly, though the cold war ensured that it remained less than planetary. The shredding of the iron curtain and the quick collapse of the Soviet Union in the early 1990s, along with the subsequent emergence of China as an economic superpower in the global marketplace, signaled that the last major obstacles to economic globalization had been hurdled. Global economic interactions exploded in the last decade of the twentieth century and have not looked back since.

Friedman suggests that the hostility between nations that kept global-ization at bay for decades during the cold war may have forever disap-peared owing to the emergence of planetwide economic relationships. Countries that are part of the same global supply chain of goods and ser-vices, he argues, do not war against each other. Friedman observes that "people embedded in major global supply chains don't want to fight old-time wars anymore. They want to make just-in-time deliveries of goods and services—and enjoy the rising standards of living that come with that."[21] Friedman's thesis is an extension of Kant's observations of the benefits of trade in the eighteenth century. It builds on the "democratic peace" argument: economic interdependence, in addition to and perhaps more powerfully than political interdependence, prevents war.

Friedman has employed the metaphor of a flattened world to describe the ever-increasing levels of global economic interdependence. He high-lights transnational business operations that spread customers, workloads, and supply chains across multiple continents. Computer systems and the Internet, in turn, have made financial interactions, trade, and sales a truly planetary phenomenon. Of course, it is not only at the supply side of economic relations that globalization occurs. Media and advertising—through radio, television, film, print, and the World Wide Web—have made the demand side of the economy equally global. Both the production and the consumption of goods and services today operate at a planetary scale. Friedman quotes Karl Marx's and Friedrich Engels's famous descrip-tion of the impact of capitalism as it spans the globe: "In place of the old wants, satisfied by the production of the country, we find new wants, requiring for their satisfaction the products of distant lands and climes. In place of the old local and national seclusion and self-sufficiency, we have intercourse in every direction, universal inter-dependence of nations."[22] Data confirm the Marx-Engels-Friedman thesis: the integration and interdependence of economic relations across the globe, with very few exceptions, have been steadily growing.[23] This economic integration and interdependence occurs between nations, between businesses, between individual consumers and producers, and between economic sectors. Eco-nomically we have never before been more connected.

Ever more connected, that is, but nobody is in charge. Of course, some of us "nobodies" are less in charge than others. And those who think they are in charge—owing to their disproportionate wealth and power—can undermine the development of sustainable societies. Still, like the natural environment, the economy is a complex, adaptive system. Like an ecosystem, the economic habitat is a multilayered, highly interactive web of relationships. In complex, adaptive systems, "individual particles, parts, or agents interact, process information, learn, and adapt their behavior to changing conditions."[24] As such, they are not limited to linear cause-and-effect relationships, such as the mechanical collisions of billiard balls on a table. Rather, economies, like ecosystems, demonstrate nonlinear dynamics: parts of the system, or the entire system, can be transformed or break down in unpredictable fashions. In chapter 1, we examined such phenomena in the natural world as cascade effects, discontinuities, and negative synergies were observed to pitch entire ecosystems into turmoil. The same thing can and does occur in the economic world, well demonstrated in the autumn of 2008 when a credit crisis on Wall Street threw the world into an economic tailspin.

Economic interdependence, like ecological interdependence, is a fact of life. Without doubt, it has its downside, but it produces many benefits as well. Studies have demonstrated that increased trade heightens productivity and industrial efficiency. It also improves standards of living, at least as measured by increasing GDP. The most economically globalized countries tend to be the most productive and wealthy; they also often have the most environmental safeguards in place. In turn, economic globalization may stimulate international cooperation, including collaborative efforts to protect the environment. Beyond environmental issues, economic interdependence may foster more cooperative and less hierarchical relationships in organizations. The business world, Friedman argues, has moved "from a primarily vertical—*command and control*—system for creating value to a more horizontal—*connect and collaborate*—value-creation model."[25]

Then there is the downside. While globalization increases global wealth, it does not distribute this wealth equitably. Far from it. When Adlai Stevenson made his final speech to the United Nations in 1965, the ten richest countries in the world were about thirty times as wealthy as the ten poorest. Today, after four decades of unprecedented economic globalization, the ten richest counties are over sixty times as wealthy as the ten poorest.[26] In addition, economic wealth within nations has not been equitably distributed in the wake of globalization. Again, inequities have grown. Today the wealthiest 2 percent of the world's citizens own more than half of the world's household wealth. Meanwhile, half of the world's population, over 3 billion people, own just 1 percent of global household wealth. In the United States, one of the most globalized economies, inequality in wealth has also increased. While the United States boasts over half of the world's billionaires, it also has the most unequal distribution of wealth of any of the world's industrialized nations and greater inequality of wealth than dozens of the world's developing countries.[27] Some degree of economic inequality is inevitable in a free society, but great disparities in wealth, political scientists since Aristotle have observed, tear apart the social fabric.

Economic globalization often stimulates international cooperation to protect the environment. But typically it does so in response to the very problems it has created or exacerbated (such as international trafficking in endangered species). And while the most globalized economies often have the strongest domestic environmental safeguards in place, they also tend to produce the most pollution, including greenhouse gases.[28] Increasing wealth and increasing concern for environmental protection generally go hand in hand. However, wealth in today's globalized economy is primarily generated through the unsustainable exploitation of natural resources and produces unsustainable levels of consumption and waste. Globalization generates wealth. And higher standards of living stimulate environmental concern. But such heightened concern is not sufficiently translated into action to offset the environmental destruction caused by the pursuit of wealth in a globalized economy.

Economic globalization has also led to the growth and concentration of corporate power. For all the increased connection and collaboration that may occur as a result of economic interdependence, there is no shortage of instances where transnational corporations have subverted and dominated local and national economies and political institutions. While economic globalization has empowered many consumers by increasing their choices in the marketplace, these same consumers, understood as citizens and workers, are often disempowered. Financial capital and business corporations easily flow across geographical and political boundaries. The labor force is not so transportable. That makes workers much more dependent on capital and corporations than the latter is on workers. Businesses today can easily relocate to cheaper labor markets in other countries if local unions, wage pressures, or environmental regulations cut into their profit margins. In turn, the institutions designed to foster free trade in a globalized economy effectively strip from local and national political bodies the ability to protect or cultivate ecological and cultural resources. The public protests that dog the annual meetings of the World Trade Organization (WTO) give evidence of citizens' concerns that they have lost control of their lives, livelihoods, and natural environments owing to the power of corporations in a globalized economy.

Corporate power also has domestic impacts. Consider a new clause of Canada's Agreement on Internal Trade that allows corporations to appeal to a dispute resolution panel if they believe local governments are restricting their operations or jeopardizing their profits. The panel can hand out fines as high as $5 million to any municipal, regional, or provincial government that restricts business opportunities. A municipality that wanted to outlaw pesticide use, or promote locally grown food in their schools, or restrict the displacement of small businesses, for example, could now be sued by pesticide corporations, chain supermarkets, or big-box stores. The trade agreement causes local governments to shy away from enacting any laws that business interests might challenge in costly court cases. As Corky Evans, a provincial legislator from British Columbia observed, "It's ceding power to corporations that you used to give to your mayor and council. . . . Ultimately, people have to realize that the old struggle between the left and the right has changed and now it's between the community and the corporate sector."[29] To the extent that globalization leaves citizens and workers with fewer opportunities to shape their collective lives, it threatens democracy, cultural preservation, and local caretaking of the environment.

Global economic interdependence is a mixed blessing. The integration of formerly isolated economies into the global marketplace provides an illustrative example. Since the 1970s, one of the key goals of U.S. foreign policy has been the integration of China into the global market economy. The transformation of China from a relatively isolated communist regime with no free enterprise and little trade into one of the world's economic powerhouses with exports whose dollar value now surpasses those of the United States marks a stunning victory for economic liberalization. But the environmental costs of China's full-fledged entry into the global marketplace have climbed at the same rate as its economic production.

China now uses half of the world's concrete and steel, is the leading importer of wood (much of it logged illegally and in unregulated forests), and has become the world's largest consumer of many other basic resources, such as grain, fertilizer, and coal. China builds a new coal-fired power plant every few weeks and uses more coal than the United States, Russia, and India combined, the next three highest-ranking consumers of this highly polluting energy source. As a result, China has surpassed the United States as the world's largest contributor to global carbon dioxide emissions. China also throws out a third of the world's garbage. The figure would be much higher if the tonnage of China's exported "durable goods" that end up in other nations' landfills were included in the tally.[30]

George Bernard Shaw once said that "there are two tragedies in life. One is to lose your heart's desire. The other is to gain it."[31] Integrating the most populous country in the world into the global marketplace did not just erode communist ideology and expand the domain of free enterprise, as American officials hoped. It also vastly accelerated the unsustainable loss of the planet's natural resources and growth of pollution and waste. The ecological shadow that China now casts is larger than that of any other country, and it will likely retain that dubious distinction for the foreseeable future. The darkening ecological shadows of other quickly accelerating economies—such as Brazil, India, and Russia, whose combined economic output is predicted to eclipse that of the world's richest nations before midcentury—also loom on the horizon.[32]

For many developing nations, becoming integrated into the global marketplace through trade has been a lifeline to increased prosperity. It has allowed export-based economic growth, a vast increase in the level of domestic consumption, and more accountable, liberalized political life. It has also led to the increased power of transnational corporations, diminishing national and local control over economic life and cultural values, and increased pressure on the planet's resources. And we must remember that sixty years of near continuous growth in trade have left more than 1 billion people living in extreme poverty, unable to meet their most basic needs of water, food, shelter, sanitation, and health care. For every $100 in economic growth as measured by GDP, it is estimated that only 60 cents (0.6 percent) contributes to the reduction of poverty for the world's poor.[33] If the growing economies of developing nations cast long and dark ecological shadows, the responsibility for lightening and shortening those shadows rests in part with the international consumers of the things they produce. When we buy goods in the global marketplace—no less than when we produce or sell them—we are never doing just one thing.

Ecological Economics

Analyzed solely in terms of GDP, the rise of the economies of China and other developing nations is wholly positive. More countries with higher GDPs means more global wealth. In this respect, the world has never before been wealthier than it is today.

GDP, defined as the market value of all goods and services that a country produces in one year, is the most widely accepted measure of a country's economic performance and national welfare. Although GDP well calculates the monetary value of market transactions, it disregards with equal thoroughness the environmental and social impact of these events. It measures everything bought and sold in the marketplace, but does not tabulate the depletion of natural resources or the pollution, social disruptions, and human costs associated with economic activity. GDP leaves out of its count a great deal that really counts.

For example, environmental catastrophes often increase a nation's GDP, sometimes substantially. The 1989 *Exxon Valdez* disaster, which dumped 11 million gallons of oil into Alaska's pristine coastal waters, had this effect. While some fishermen were thrown out of work, the massive cleanup efforts, duly recorded as market transactions, produced a net gain of economic activity. In the same vein, personal or collective efforts to lessen the environmental impact of our lives often decrease GDP. Walking, biking, or taking mass transit to work contributes less to GDP than driving an automobile. Wearing a sweater on a cold winter night contributes less to GDP than blasting the furnace with the windows open. In general, the environmental dictum to "reduce, reuse, and recycle" is a drag on a nation's economic performance measured by GDP. Likewise, manufacturing weapons for deadly wars contributes more to GDP than living in peace. Expensive, drawn-out divorces increase GDP more than happy marriages do, especially if the children of these broken families require costly, long-term psychological therapy.

In the early 1970s, E. F. Schumacher had already observed that national economic systems employed a funny way of measuring things. When natural resources were exploited, national accountants tallied this on the plus side as a form of economic production. In other words, they treated the using up of "natural capital"—the stock of resources supplied by the earth—as if it were income. Emitted pollution, however, did not find its way into the accounting books at all.

The capacity of the biosphere to absorb pollution—what Schumacher calls "the tolerance margins of nature"—is just another form of natural capital.[34] Once this tolerance margin is surpassed, a normally renewable resource ceases to replenish itself. A wetland's capacity to absorb pollution from runoff, for instance, might continue indefinitely so long as it remains a vibrant ecosystem. But if runoff pollution exceeds the tolerance margin of the wetland, killing off all or most of its flora and fauna, it will no longer absorb and recycle pollutants. The wetland will become a wasteland. The saturation of the land, air, or sea with pollution constitutes an unsustainable depletion of a natural resource. When nonrenewable resources are exploited and renewable resources are pushed beyond their tolerance margins, the economic activities in question should not be tallied in the black. The "modern industrial system," Schumacher stated, "lives on irreplaceable capital which it treats as income."[35] Its accounting books are being fudged.

Schumacher's concerns were shared by economist Herman Daly, whose work prompted the development of ecological economics as a field of study. Ecological economics, which examines the interdependence between the economic sphere and the biosphere, is founded on the conviction that the economy is a "wholly owned subsidiary" of the environment. Along with Schumacher, Daly insists that we "stop counting the consumption of natural capital as income."[36] He argues for a "steady-state economy" that operates in a bounded fashion, as a "physical subsystem of the ecosystem." The man-made or anthropogenic capital of technology, human ingenuity, and social organization, Daly observes, can complement the natural capital provided by ecosystems. But anthropogenic capital cannot replace natural capital and should not be confused with it.[37] Whereas man-made capital is potentially limitless, natural capital is a finite resource. The dream of endless economic growth, which animates most of the world's political economies and is hawked to constituents by virtually every politician, is just that—a fantasy. The finite boundaries of natural capital ensure that the dream of endless growth will turn into a nightmare of environmental collapse.

Daly writes, "The scale of the economy must remain below the capacity of the ecosystem sustainably to supply services such as photosynthesis, pollination, purification of air and water, maintenance of climate, filtering of excessive ultraviolet radiation, recycling of wastes, etc."[38] The services that nature supplies are massive but finite. Attempts have been made to estimate the dollar value of the planet's ecosystem services, such as pollination, nutrient cycling, and flood control, to name but three of a long list.[39] Efforts to measure natural capital in monetary terms are inevitably speculative and inaccurate, and they have the potential to demean nature's bounty by reducing it to a dollar figure. But they confirm that human economic production remains dependent on the vastly larger productivity of nature.

To remain within the limits set by the planet's natural capital, human economies cannot grow indefinitely. Or rather, their growth cannot be based on increasing exploitation of natural resources and rising emissions of pollutants. As Kenneth Boulding reputedly observed, to believe that endless growth is possible in a finite world, you have to be either a madman or an economist. Consider, with this in mind, the oft-stated argument that we require population growth to safeguard social security. More youthful wage earners are needed so that more elderly retirees can receive their benefits. Effectively, what is being promoted is a giant Ponzi scheme.[40] Ponzi schemes inevitably collapse, and many people will lose their shirts. In Seussian terms, biggering cannot go on forever.

Certain forms of growth, it is fair to say, are potentially endless. Social and cultural growth, perhaps better termed development, is a case in point. Social and cultural relationships can always be enriched and enhanced. Technological innovation as well may be continuous. Endless and even exponential growth in these realms is possible, and often beneficial. The same cannot be said for that portion of economic growth grounded in the depletion of natural resources. As Schumacher observed, human needs are infinite, and infinite needs can never be satisfied with material goods.[41] The planet's capacity to provide raw materials for, and absorb waste from, our systems of economic production and consumption is very large, but it is clearly limited.

Economic systems that ignore such limits are not counting what really counts. The problem stems from what economists call "externalities." Businesses often engage in the practice of externalizing some of the environmental and social costs of production. For example, a pulp and paper mill in pursuit of profit must ensure that the side of the ledger listing the costs of production—primarily the raw material of wood pulp, water, and chemicals; the energy employed at the mill; machinery purchases and maintenance; and labor—tallies up less than the side of the ledger listing revenues from sales. Otherwise the mill cannot stay in business. With this in mind, the mill may dump the wastewater produced by its operations into a nearby river. Effectively it passes along the costs of cleaning up the polluted waterway, and the risks associated with it, to residents living downstream.

Externalizing costs means fobbing them off on others. The fobbed-off cost may come in the form of real expenses that have to be paid, such as the expense borne by downstreamers when they clean up a polluted river. The fobbed-off cost may also come in the form of a (potential) benefit never received, such as a pristine river that downstreamers cannot afford to maintain. Depleting nonrenewable natural resources may also be viewed as externalizing costs. In this case, the cost is fobbed off not to geographical downstreamers but to temporal downstreamers. It is an ecological and economic deficit passed on to future generations.

Externalizing the costs of doing business is a way of making financial gain by fudging the books. A recent United Nations report estimates that one-third of corporate profits are gained by externalizing environmental costs.[42] To get the accounting right, ecological economists recommend full-cost pricing. Here the prices that businesses charge for their goods and services factor in the long-term environmental and social costs associated with the production, delivery, use, and disposal of products. Full-cost pricing, also known as full-cost accounting, eliminates externalities.

Programs for extended producer responsibility go some distance to full-cost pricing by expanding the liability of manufacturers and distributors of consumer goods for the waste streams they produce. About half of the states in the United States have enacted legislation involving extended producer responsibility for electronic waste.[43] Some, if not all, of the cost of recycling electronic waste in these states is borne by industry and passed on to consumers or directly levied on consumers in either added fees or in the purchase price for electronic goods. So-called green taxes are another step toward full-cost pricing. Taxes on gasoline, for instance, may be employed to cover the costs of building and maintaining highways;

defraying police, ambulance, and medical services associated with vehicular traffic and accidents; and addressing the various forms of pollution, including greenhouse gas emissions, caused by vehicle exhausts. In many countries, the automobile remains the beneficiary of a massive welfare program. Its use is significantly subsidized by governments, which get (mostly local and state) taxpayers to pony up for the externalized environmental and social costs associated with the transportation system and the burning of fossil fuels. Were the price of gas in the United States to reflect the full cost of automobile use, it would likely triple or quadruple.

Likewise, many agribusinesses externalize the environmental costs of their operations, including the depletion of freshwater sources, soil erosion, the eutrophication or nutrient loading of water sources caused by fertilizer use and runoff, and the accumulation of contaminants caused by the extensive use of pesticides and herbicides. In turn, the environmental costs of the delivery of agricultural products to consumers are also externalized, including the pollution generated and energy expended by extensive transport and refrigerated storage. Agribusiness today is highly profitable. If the environmental costs of doing business could no longer be externalized, however, it is estimated that most agricultural producers would be operating at a substantial loss.[44]

Cambridge University economist Arthur Cecil Pigou argued in the 1920s that governments could "internalize the externalities" of the market through subsidies and taxes that deliver welfare programs. Full-cost pricing makes this internalization of externalities transparent and more direct. It ensures that the businesses and industries responsible for the externalities themselves bear the cost of internalizing them, with these costs passed on to the consumers of their products and services. Apart from ensuring fair and accurate economic accounting, full-cost pricing encourages efficient resource use and the development of green technology. It would also stimulate local economies, free up governmental resources and revenues for other public goods, and produce fewer social and environmental casualties.

Unfortunately, many macroeconomic policies are directly opposed to full-cost pricing. Worldwide, governments spend $1.5 trillion a year in subsidies, about half of which, $700 billion, contributes to environmental degradation. American families pay well over $1,000 annually in subsidies to businesses, many of which are environmentally destructive.[45] Between

2008 and 2013, the oil and gas industry alone was slated to receive over $30 billion in direct tax breaks and other subsidies.[46] Worldwide, hundreds of billions of dollars are spent annually subsidizing fossil fuel production and use.[47] This is corporate welfare, not a free market.

In his promotion of a free market, and opposition to the protectionist and trade-hindering mercantilism of his day, Adam Smith wrote that "consumption is the sole end and purpose of all production; and the interest of the producer ought to be attended to, only so far as it may be necessary for promoting that of the consumer."[48] Corporate welfare is not in the interest of consumers. It occurs in the form of direct government subsidies to industry and through the indirect subsidies that result when public coffers are drained to remedy the costs that businesses externalize to society. Adam Smith, the founding father of capitalism, was not in favor of it.

In many ways, however, we have adopted Smith's recommendation for holding the interests of consumers foremost. Consumption has become the mainspring of the economy and, to a significant extent, a way of life. In a 1959 edition of *The New York Journal of Retailing*, Victor Lebow wrote: "Our enormously productive economy . . . demands that we make consumption our way of life, that we convert the buying and use of goods into rituals, that we seek our spiritual satisfaction, our ego satisfaction, in consumption. . . . We need things consumed, burned up, worn out, replaced and discarded at an ever increasing rate."[49] Lebow's recommendations have largely been adopted, and much economic growth has resulted. But devastating costs are associated with resource-depleting, consumption-driven economies. Adam Smith notwithstanding, neither consumption nor production has a "sole end." There are always unintended consequences.

With this in mind, a British report on the economic drivers of global warming rightly observed that climate change constitutes a "market failure" of the greatest scope.[50] Oystein Dahle, former Exxon vice president for Norway and the North Sea, put the broader point succinctly: "Socialism collapsed because it did not allow prices to tell the economic truth. Capitalism may collapse because it does not allow prices to tell the ecological truth."[51] The truth that Dahle refers to are the ignored or hidden externalities—the unintended consequences of our economic activities. Ecological economics attempts to measure what matters and get the price right. The goal is to avoid market failure by helping the market tell the truth. Key to this effort is heightened transparency.

Transparency in business operations is a crucial means for citizens and consumers to inform their own civic engagement and market activities so they may better protect themselves and pursue their interests. The idea is that citizens and consumers have a right to know what social and environmental impacts the products they buy, and the business operations that share their *oikos*, have on their local communities, their nations, and the planet. Rachel Carson's seminal investigations of the effects of biocides stimulated the development of right-to-know legislation that forced businesses to disclose the chemicals, contaminants, and pollutants they were producing and emitting. These data, now easily available on the Internet, allow people to inform themselves about the safety of their workplaces and neighborhoods.[52] Public information sites and ecolabeling programs that reveal the social and ecological and health impacts of manufactured products also contribute greatly to transparency.[53]

Counting and Discounting

Gross domestic product neither tells the whole economic truth nor offers an accurate means of measuring social and environmental well-being. Schumacher's and Daly's seminal works have stimulated a number of efforts to develop alternative means of quantifying quality of life at regional, national and global levels.[54] The Happy Planet Index (HPI) was developed by the New Economics Foundation (whose motto, "Economics as if People and the Planet Mattered," is an expansion of the subtitle of Schumacher's *Small Is Beautiful*). The most recent HPI (2.0, based on 2005 data) ranks 143 countries according to three combined indicators: high life satisfaction (measured by survey responses), high life expectancy (gleaned from national statistics), and per capita ecological footprint (determined by the amount of land required to provide for all the resources a citizen of a particular country consumes, plus the amount of vegetated land that would be required to absorb carbon emissions produced directly by the individual or embodied in the products he or she consumes).[55] Canada's Sustainability Indicators Initiative is one of the better examples of a national effort to gauge progress toward social and environmental health and welfare.[56] The small Himalayan nation of Bhutan, as well as many

European states and other countries across the globe, are actively developing the means to replace or supplement GDP with other sets of indicators that more closely assess GNH—gross national happiness.

Well-being and happiness are difficult values to define. Devising and evaluating indicators that accurately capture these values is even more difficult.[57] Still the effort to gauge quality of life rather than GDP goes a long way to measuring what matters. Ultimately, however, to get prices in the marketplace to tell the truth requires full-cost accounting. It is seldom practiced for two primary reasons: it threatens profitability (businesses that externalize costs can enjoy higher profit margins), and it is an extraordinarily challenging task.

To be done well, full-cost accounting entails a life cycle assessment of the social and environmental impacts of products and services from cradle to grave. Life cycle assessment begins with the extraction and use of raw materials that go into products. It further investigates the impacts of the manufacturing processes, examines the effects of distribution, and assesses how the product is used and disposed. In examining these phases in the life cycle of products, analysts must attend to the production of toxic wastes and pollution, including greenhouse gas emissions, the extent of natural resource depletion, habitat destruction and land degradation, as well as an extensive list of potential social impacts.

To a very limited degree, such assessment and accounting already occurs under the guise of cost-benefit analysis, also known as risk-cost-benefit analysis. Here the risks and costs of any proposed business activity or transaction are weighed against expected benefits. The activity in question might be the development of a new product, the introduction of a new technological process, or, alternatively, the introduction of a regulation, restriction, or tax on a current or proposed product or process. The goal is to translate potential risks into monetary costs, add these to direct monetary costs associated with the activity, and compare the combined total with the expected benefits, again rendered in monetary terms. Theoretically the balance sheet should yield clear answers captured in dollars and cents.

Lord Kelvin stated, "When you can measure what you are speaking of and express it in numbers, you know that on which you are discoursing, but when you cannot measure it and express it in numbers, your knowledge is of very meager and unsatisfactory kind."[58] Kelvin was addressing the merits of scientific research that employed rigorous methodologies

grounded in the scrupulous collection of empirical data. Cost-benefit analysis and, even more so, full-cost accounting also require gathering data
and accurately tallying numbers. The problem is that these data and
numbers are not always easy to come by. Not everything of worth can be
counted or counted well.

Cost-benefit analysis translates all goods into a monetary value. It
assumes that everything has a price. Imagine a new product (e.g., a food
additive, technological innovation, or industrial process) that will prove
profitable in the marketplace but will marginally increase the occurrence
of a particular disease among its consumers. Cost-benefit analysis would
have us weigh the benefits of introducing the new product, measured in
profits, against risks and costs. Risks and costs would be quantified by
multiplying the increased probability of disease by the costs of medical
care to treat the disease and the costs of human resources lost to sickness
and early death. Do the math, and you have your recommendation.
However, medically tending to disease and monetarily compensating
bereaved kin does not produce the same level of human welfare as does
maintaining healthy consumers and citizens. Cost-benefit analysis fails to
account for this fact.

Cost-benefit analysis also fails to address the rights of those who may
be harmed. As a member of the Natural Resources Defense Council
observed, standard mechanisms of economic analysis "sidestep" basic
issues of justice: "Imagine a law requiring federal agencies to measure the
worth, in dollars, of free speech or civil rights before acting to enforce
them: any attempt to weigh an intangible public good against hard cur-
rency will shortchange the former, whether it is the right to vote or the
value of protecting a human life against cancer."[59] Not all public goods are
"intangible." Human health is relatively easy to define and assess, but
human health and many other goods such as beauty, friendship, and
human rights cannot be reduced to monetary values without undermining
their intrinsic worth.

The problem here is that cost-benefit analysis requires a standard metric
for all its evaluations. Reducing a good to its equivalent in dollars and
cents, however, fails to capture all that exceeds or defies its marketability.
In other words, the market value of something is not its full value. Much
of what we value in life is "trans-economic."[60] This is certainly true for
goods like friendship, love, and human rights, which we immediately

understand to be priceless. But many natural resources that are bought and sold in the marketplace also defy accurate pricing. Carl McDaniel observes:

By setting a price for a natural resource, we presume that we know its total value, which in many cases is certainly not true. . . . For example, what should be the price of the Catskill watershed? What value should we ascribe to its clean water supplied to New York City? Its agricultural output? Its recreational use? Its aesthetic value? Its forests and the carbon they sequester from the atmosphere or the flood control they provide? Or the still-unidentified life support it provides? And how shall we calculate the "correct" price for all of these "goods and services" and those that future generations will need but are still unknown? Clearly preservation of some natural resources will be enhanced through market mechanisms, but many will not. In fact, many natural resources will be harmed, diminished, or eliminated by market forces. The challenge is to identify and appropriately price those elements of natural resources that the market can efficiently allocate while creating mechanisms that preserve those aspects of the natural world that have little or no currently accepted value yet may have unrecognized present or future value.[61]

With this in mind, we might refuse to "subordinate the very creative force of the universe to a[n economic] system that is simply one small product of that force."[62] The value of wilderness, like many other goods, cannot be wholly translated into monetary terms without, by that very translation, degrading it. Nature has economic value; but it is worth much more.

Cost-benefit analysis is a notoriously imprecise venture.[63] The problem only gets worse when aesthetic, spiritual, and ecological costs, risks, and benefits are thrown into the equation. The difficulty is heightened when calculations have to be extrapolated to account for future generations. At the end of the day, however, economics is about numbers. We may always refuse to assess the value of a (public) good for fear of miscalculating or degrading it. But things that are not measured will not be counted. And what is left uncounted will seldom be preserved. In all likelihood, it will become an externality.

With concern for such externalities, Thomas Friedman advocates "Market to Mother Nature" accounting:

It's now obvious that the reason we're experiencing a simultaneous meltdown in the financial system and the climate system is because we have been mispricing risk in both arenas—producing a huge excess of both toxic assets and toxic air that now threatens the stability of the whole planet. Just as A.I.G. sold insurance derivatives at prices that did not reflect the real costs and the real risks of massive defaults (for which we the taxpayers ended up paying the difference), oil companies, coal companies and electric utilities today are selling energy products at prices that do not reflect the real costs to the environment and real risks of disruptive climate change (so future taxpayers will end up paying the difference).[64]

Friedman's linkage between financial collapse and ecological collapse is astute. To argue the need for full-cost (or at least more accurate) pricing is not to undermine the free market. Quite the opposite is true. Full-cost pricing is an effort to make the market work more efficiently and transparently. The alternative is more government intervention as billion-dollar bailouts force taxpayers to shoulder the burden of market failures.

The economic market is reputedly a place of hard-nosed realism. Yet economists often forget a fundamental tenet of realism: you cannot get something for nothing. If you don't leave cash on the table for your meal, somebody, somewhere, at some point, will have to foot the bill. There is no such thing as a free lunch. This is one of Barry Commoner's four laws of ecology. To externalize costs is to promote free lunches.

Consider the unintended consequences of literally providing free lunches. In times of famine, desperate nations may receive food aid from other countries. The underlying problem often is not that the afflicted nations cannot grow enough food for themselves. The problem is that the best land is used not for domestic food crops but for exported cash crops, and the food that is grown is ineffectively distributed. When famine strikes these countries, donor nations effectively subsidize their own agribusiness to grow food that is given away overseas. The agribusiness from donor nations does quite well as a result. But local farmers in afflicted nations may suffer. Their market is effectively undercut. Faced with the distribution of free or highly subsidized imported food, local farmers in developing nations can be put out of business. Afflicted nations then become more dependent on imported food even in nonfamine years. As a result, a cycle of agricultural weakness, periodic famine, and foreign dependence is created.[65] This is not to gainsay the merits of aid, especially aid that supplies basic needs to desperate people. But you can never do just one thing, even when you are being charitable.

Consider another example of the perverse effects of free lunches. The economic collapse of 2008 was caused by the pursuit of free lunches by a financial sector engaged in gambling. Gambling appears to create wealth out of nothing, or nothing more than the willingness to assume risk. But somebody, somewhere must pay for this newly created wealth. In casinos, there are always at least as many losers (and usually many more) as there are winners. When Wall Street hedge fund managers and derivative traders grossed hundreds of millions of dollars in annual incomes for gambling with highly risky mortgages, it seemed that large amounts of wealth were being created out of thin air. Everyone could be a winner. But every financial gain in the absence of real productivity constitutes a loss somewhere by somebody at some (future) time. When the house of cards came crashing down in fall 2008, it became clear how numerous the losers were. Beyond those families who had to forfeit their overpriced and overleveraged homes, every taxpayer in the United States became a big loser, as did the 11 million Americans who lost their jobs in the ensuing economic recession.

There is no such thing as a free lunch. Living on credit, however, would have us believe otherwise. When credit card companies, in the 1980s, began offering no annual fees and cheap (if any) interest in their introductory offers to consumers, it might have seemed too good to be true. And it was. Few people read, or were capable of understanding, the very fine print written in these offers. Penalty fees and escalating interest rates soon left countless individuals and families with growing debts they could not repay. Living on credit defers the true cost of our economic activities. The future may whisper to indebted individuals. "One day," it softly murmurs, "the piper will have to be paid." But the present shouts too loudly for this message of prudence to be heard.[66] The costs of goods consumed get externalized to a future without voice.

Economists assume that a "positive time preference" constitutes a basic feature of human nature. Losses that we will suffer tomorrow, they know, are not feared as much as losses we will suffer today. Benefits enjoyed immediately are valued more than those that must be deferred. In the jargon of economic theory, the future gets "discounted." Discounting the future is a natural human tendency. A bird in the hand, after all, is worth two in the bush. Money that we have in our possession is secure and can be invested for a profit. Discounting the future makes economic sense. What makes economic sense, however, may not make ecological sense.

Within financial systems, gains or losses that will not be realized for decades are often discounted to zero.[67] At such discount rates, as John Dryzek observes, "a system may be judged economically rational while simultaneously engaging in the wholesale destruction of nature, or even, ultimately, in the total extinction of the human race."[68] A discounted future is a bleak place. When we discount the future at a high rate, we are effectively saying that progeny cannot count on us.

Debt, Throughput, and Sustainable Economies

As a species, we are living beyond our means, and the economy of nature is tanking. Natural capital is being depleted at an accelerating rate, hastening the day of biospheric bankruptcy. In many respects, our indebtedness in the natural economy parallels our indebtedness in household and national economies. For much of the previous decade, almost half of American families spent more each year than they earned. Consumer debt, which does not include home mortgages, nearly doubled in the ten years preceding the financial collapse of 2008. Personal debts accumulated to $8,000 for every man, woman, and child in the nation, a figure that must be quadrupled to capture the per capita portion of the national debt. This public debt is what the U.S. government now owes various domestic and, increasingly, foreign creditors on behalf of its citizens.

While personal debts are eventually paid by the people who incur them (unless they declare bankruptcy), that is not the case with national debt. Much of this public debt will be passed on to future generations. As a percentage of its GDP, the United States ranks in the top quarter of all nations in terms of the size of its national debt. It borrows over half of all the money lent to the world's nations. In 2010, the federal government ran a deficit of over $1.5 trillion, setting a post–World War II record. That is the amount of money it had to borrow in one year simply to stay solvent. The figure is more than 10 percent of the total amount of goods and services (GDP) that the nation produced. For every $1,000 that exchanged hands in 2010, another $100 had to be borrowed to make ends meet. Federal deficits are predicted for each year of the coming decade.[69]

The United States is not the only country currently in financial hock. Across the globe, nations are mortgaging the financial future of their youngest and yet-to-be-born citizens, just as the species as a whole is mortgaging its ecological future. Edmund Burke insisted that current generations must be mindful of "what is due to their posterity." They have no right to squander their inheritance, passing on to future generations a "ruin" rather than a "habitation."[70] With an eye to fiscal matters, both George Washington and Thomas Jefferson maintained that each generation must pay its own debts. The failure to do so on a national scale, they believed, burdened posterity with deprivation and the threat of war. Today, however, we frequently buy what we cannot afford. As individuals, nations, and a species, we are living beyond our means. Depleting our natural and financial capital, even to the point of being significantly in hock, has become something of a norm.

When we incur debt or deplete capital in other ways, we are discounting the future. Discounting the future at high rates makes economic sense only if resources are infinite and economic growth unending. Only in such a system might a discounted future regain its value by forever increasing the amount of economic activity it harbors. In such an infinite system, individuals or nations might reasonably expect to "grow out" of debt. In fact, our world is not infinite.

Herman Daly develops the notion of throughput to underline the fundamentals of economics in a finite system. *Throughput* refers to the total quantity of input and output of economic activity. It includes the input from extracted raw materials and from the energy used in the manufacturing, transport, and marketing of goods. It also includes the waste outputs of every phase of this process, from raw material extraction to postconsumer disposal. "Throughput begins with depletion," Daly writes, "and ends with pollution."[71] In the United States today, the sheer size of the material throughput is mindboggling. It amounts to over twenty times each citizen's body weight per day, which is to say, over 1 million pounds for every man, woman, and child annually.[72] Most of this throughput is industrial waste. Indeed, only 6 percent of materials mined, extracted, or harvested actually end up in the products that we buy and sell.[73]

Every economy has throughput. However, a sustainable economy restricts throughput to levels that do not tax the earth's regenerative capacities beyond their limits. In turn, it compensates for the depletion of nonrenewable resources. Those who produce pollution or deplete natural capital are held responsible to offset these degradations or depletions.[74] To this end, "polluter pays" and "depleter pays" principles form the foundations of a full-cost accounting. The cost of pollution abatement is relatively easy to determine. In the case of the use of nonrenewable natural resources, depleters must bear the cost of the development of renewable resources.[75] Taxes on fossil fuel extraction and use, for example, might be used to pay for planting trees to offset emissions and the development of renewable energy sources to ensure clean, abundant energy for the future.

The costs of such offsets ultimately would be passed on to consumers of the goods and services that produce the depletion or degradation of natural resources. Goods and services that do not rely on high levels of throughput therefore would be at a competitive advantage. In such an economy, one would expect the accelerated expansion of green technology and other forms of economic development that provide goods and services with diminished throughputs. Daly's term *steady-state economy* may mislead in this respect. A sustainable economy is anything but moribund. While spurning ever-increasing throughput, a sustainable economy depends on and stimulates innovation and human ingenuity.

Nature does not make permanent solutions. It simply evolves organisms and ecosystems that work well in particular contexts. Once evolutionary change stops, life stops. A sustainable economy models itself on nature. It does not produce a permanent solution to the challenge of satisfying human wants and needs, for these wants and needs change with time and place, as do the habitats within which they can be satisfied. To develop a sustainable economy, we must become agents of change, not simply because we must alter the way we currently do business but because sustainability is in the business of change. As ecologists observe, to be sustainable, we must

identify and reduce destructive constraints and inhibitions on change, such as perverse subsidies. Protect and preserve the accumulated experience on which change will be based. Stimulate innovation in a variety of safe-to-fail experiments that probe possible directions, in a way that is low in costs for people's careers and organizations' budgets. Encourage new foundations for renewal that build and sustain the capacity of people, economies, and nature for dealing with change. Encourage new foundations to consolidate and expand understanding of change.[76]

To be sustainable, an economy must be resilient, and to be resilient, it must be adaptive.

In a sustainable economy, change occurs primarily through the innovation and industry of a free market where producers tell and consumers learn the ecological truth of their transactions. Here there are no fail-safe solutions. Rather, there are lots of safe-to-fail experiments. These experiments—the development of new products, new forms of economic organization and cooperation, new technological processes and inventions, new ways of offsetting and compensating for the throughput we cannot avoid—are adaptive mechanisms. Some will succeed and lead to best practices and enduring value. Many will fail. But their failures will not lead to economic or ecological collapse. Rather, they will stimulate more adaptation. John Dewey, the early twentieth-century American philosopher and social theorist, rejected the task of creating a planned society. Instead, he advocated a *"continuously planning society."*[77] We might say the same about a sustainable economy. A planned economy deprives people of the benefit of entrepreneurial energy and imagination while saddling them with bureaucratic burdens. A continuously planning economy is both dynamic and future focused. A sustainable economy is not planned. It is adaptive. Its market is free without being wasteful or depletive.

The justification for spending hundreds of billions of taxpayers' dollars to bail out investment firms and banks, and subsequently industrial firms, in the wake of the market collapse of 2008–2009 was that these transnational corporations were "too big to fail." If they folded, the magnitude of the collapse, owing to their sheer size, would create an avalanche of economic destruction at home and abroad. Sustainable economies do not put all their eggs in so few baskets. Like a healthy ecosystem, they rely on the competitive and cooperative interaction of many different organisms: individual entrepreneurs, small businesses, corporations, cooperatives, and government agencies that tax and regulate when necessary to ensure that market prices come as close as possible to telling the truth. Diversity builds resilience. The dynamic interactions of multiple stakeholders ensure economic stability in the face of inevitable change.

Money, Markets, and Happiness

While discounting the future is a natural human tendency and, in many instances, makes good economic sense, like all other human tendencies and habits, it becomes perverse when taken to extremes. The capacity to live in the present moment without becoming obsessed with plans for the future is a sign of psychological and spiritual health. But living in the here and now is not an excuse for burdening progeny with debts. And we should not mistake endless consumption that we can ill afford with living in the here and now. When shopping becomes the national pastime and consumption becomes a personal obsession, we are not exercising spontaneity in the marketplace. Rather, we are fleeing the complex inter-dependence of the here and now, and the anxieties it may produce, into a dangerous diversion.

All too often, consumption in modern society is a means of escaping into the fantasy that one can buy happiness. We pretend that the next purchase will finally bring us joy, only to discover that the treadmill of consumption is accelerating and the happiness we seek remains just out of reach. The discarded goods and hopes of yesterday fly off the back of the treadmill into landfills and incinerators. And with each accelerated step toward the next purchase, we nervously sense that stopping, or even slowing, would be too risky. Consumption becomes an obsession that is demanded by and feeds the national addiction of endless high-throughput economic growth. It is a dangerous game that does not deliver welfare or happiness.

Survey research confirms that people are not happier today in the United States or other developed nations than they were in the 1950s, notwithstanding much greater levels of consumption—indeed over a 300 percent increase in per capita GDP. Dire poverty is no recipe for happiness. However, once people have met their basic needs for decent housing, food, and health care, more money and greater consumption do not yield pro-portionately greater happiness.[78] Without health, being happy is very chal-lenging. And without sufficient income to meet basic needs, it is difficult to be happy even with good health. Establishing a secure and sufficient income goes a long way to ensuring well-being and life satisfaction. But where money is concerned, more is not always better. Of course, people typically prefer to be wealthier, and they typically believe that becoming wealthier will make them happier. But they are often mistaken.

We are always comparing ourselves to others and gauging our happiness by this metric. Studies demonstrate that people will choose to be relatively richer (or better looking) than their neighbors rather than having greater absolute wealth (or better looks) within a wealthier (or better-looking) community.[79] In other words, people choose to be poorer (and uglier) than they might be as long as they stay richer (and better-looking) than their immediate neighbors. This human tendency for comparative evaluation explains why people think they will be happier if they have more money. Acquiring more wealth would make people better off than the cohort against which they been in the habit of comparing themselves. The problem is that when people actually become richer (they get a substantial raise or win the lottery, for example), a temporary euphoria quickly dissipates. Soon they find themselves back at their previous level of happiness (if that) because they now compare themselves to a new cohort of people with a higher economic status.[80] In other words, the grass will always appear greener, and our lives will promise to be happier, a little further up the mountain. So data from the social sciences buttress what sages have always counseled: money cannot buy happiness or the love and fellowship that foster the greatest happiness. Of course, a sustainable economy cannot guarantee happiness either. But it is a prophylactic against some of the fantasies that undermine its pursuit.

The alternative to overconsumption and its accompanying fantasies is not the hoarding of wealth. Miserliness is not an individual or social virtue. Money is meant to circulate. If economic activity is one of the chief vehicles we have to experience our interdependence, clearly a money-grubbing stinginess is counterproductive. For those with disposable income who take sustainability seriously, it is important to consume well (buy the right things, such as healthy foods and well-crafted, long-lasting goods that are carbon light), waste less (buy fewer waste-heavy and unnecessary things), and pay fair prices (such that producers up and down the supply chain receive living wages). The point is not to opt out of economic life but to contribute to a sustainable economy.

In no other field of human endeavor, with the possible exception of ecology, is the interdependence that structures our lives more evident than in economics. While the marketplace often teaches aggressive competitiveness, its classroom offers many other lessons. After all, the market provides a wonderful venue for connecting with other people. Cooperation is stimulated by face-to-face encounters, and notwithstanding the growth of Internet shopping, no other place rivals the market in providing an abundance of personal encounters and encouraging the voluntary cooperation of large numbers of people. Indeed, the market can and should be a place where we learn to interact with and trust other people. One of the best indicators of a country's standard of living, general economic health, and extent of investment is the level of trust shared by its citizens.[81] Markets work best when people trust each other. And the trust and cooperation that make markets work well can be cultivated within markets themselves—if the goods bought and sold there tell the truth.

Everybody in the marketplace wants to connect to someone else, if only to make a sale or purchase. And everyone in the marketplace has the opportunity to transform and deepen this connection. The conversation with the barber who cuts your hair or the cashier who tallies your purchases provides this opportunity. Economic exchanges, of course, vary greatly in their capacity for more expansive connection. But each presents an occasion to experience and explore the complex interdependence that defines our lives.

Seventeen years before *The Wealth of Nations* appeared, Adam Smith published *A Theory of Moral Sentiments*. Here the Scottish philosopher who set capitalism on its course outlined the most basic human motivations, needs, and aspirations. While a capitalist economy is grounded on the individual pursuit of self-interest, Smith realized that human beings were not only or even primarily economic actors. "How selfish whatsoever man may be supposed," Smith began his book, "there are evidently some principles in his nature, which interest him in the fortune of others, and render their happiness necessary to him, though he derives nothing from it, except the pleasure of seeing it."[82] What is central and natural to every human being is compassion, sympathy or fellow feeling. Human beings are naturally disposed to take joy in others' pleasures and suffer from their pain. As one commentator describes Smith's understanding of the role of sympathy: "In the jargon of the present day, individual utility functions are interdependent, according to Smith—often highly interdependent."[83]

Before Smith laid the groundwork for modern capitalism, sketching out the nature of its relationships, he acknowledged that human beings were psychologically interdependent. He also insisted that this emotional connection to other people, not our self-interested pursuits in the marketplace, was what ultimately made us happy. "The chief part of human happiness," Smith states, "arises from the consciousness of being beloved."[84] Empirical research confirms Smith as an astute observer of the human condition. At least, he got it half right. Our happiness depends on our sense of being loved, but also, and in equal degree, on how much we love others. Positive psychologists have found that loving and being loved are crucial components of inner joy and, more broadly speaking, happiness.[85] Our integration into an interdependent community of caring is the surest path to happiness.

Even the wealthiest of today's billionaires cannot take a penny past their graves. Self-interested pursuits inevitably are short-term successes. In contrast, the legacy of our contributions to the welfare and sustainability of a community, as we observed in chapter 2, can endure for millennia. These contributions, coupled with the love we receive and bestow on others, allow us to experience the greatest happiness. These are not idealistic speculations. They are the facts as we know them, established by social research and confirmed by personal experience.

It is said that the key to happiness is lowered expectations. Given the comparative nature of our judgments, there is something to this quip. Beyond a rather modest living, ever greater wealth and levels of consumption do not yield greater happiness. Rather, contributing to the greater good, cultivating caring relationships, and gratitude for what we have are better recipes for contentment. Of course, humans are comparative evaluators. But there are countless realms of life within which excellence may be achieved—with lots of room for accolades, pride, and self-interest—beyond economic endeavors. A sustainable economy would provide the material foundation for these varied pursuits, and the good life they allow.

Amory Lovins, the author of *Natural Capitalism*, argues that "economies are supposed to serve human ends—not the other way round. We forget at our peril that markets make a good servant, a bad master and a worse religion."[86] The blind faith that unregulated markets could save the world imploded with the financial collapse of 2008. The conviction that markets can and should externalize their social and ecological costs has been steadily eroding for fifty years. Climate change has fully exposed its perversity. Transforming the market into an excellent servant is a daunting task, but the stakes are too high for complacency and inaction.

In a sustainable economy, where transparency allows the goods and services we buy and sell to tell the ecological and social truth, the pursuit of self-interest is a good thing. It is the engine of prosperity—understood not as the endless accumulation of personal wealth, but the fullest meeting of material, social, and psychological needs in a sustainable fashion. Not all unintended consequences are bad. The goal of a sustainable economy is to ensure that the pursuit of self-interest in the marketplace, apart from serving its explicit purpose of benefiting the individual, has the unintended consequence of creating public benefits for both current and future generations.

Conclusion

The free market, in one form or another, is hundreds of thousands of years old. It began with the exchange of basic goods and services among our hominid forebears. The first market transaction might simply have been: "I'll scratch your back, if you'll scratch mine." Goods may have been limited to a few items of food or basic tools and weapons. Services were limited to nit picking, child care, tool crafting, or shelter construction. Most likely there were still only a few hundred goods and services in exchange in early agricultural societies, as recently as 8,000 to 10,000 years ago. Today the global economy presents us with goods and services numbering in the tens of billions. Our economic interdependence is broad and deep.

Contemporary market economies, like ecosystems, are complex adaptive systems. Optimally they facilitate diversity, operate through open information flows and feedback mechanisms, are largely self-regulating, prove resistant to external control, and are dynamic and nonlinear in their mechanisms.[87] With this in mind, the opportunity before us is to create an economy where the pursuit of self-interest creates wealth without increasing our vulnerability to crisis, where trust and cooperative human relationships can be cultivated in tandem with competitive interaction, where social equity is recognized as an intrinsic component of prosperity, and where constant adaptation and small-scale experimentation safeguard us from catastrophic collapse.[88] But why should we believe that such a sustainable economy is possible when the history of humankind presents an extended testimony to the shortsighted, wasteful, endless, and often tragic pursuit of wealth?

For the first time in history, we can clearly see where business as usual will lead. A globalized marketplace that does not tell the social and ecological truth of its transactions in an era of plummeting biodiversity, dwindling natural resources, and accelerating global warming threatens civilization itself. Absent ecosophic awareness, the blind pursuit of self-interest by a growing planetary population that has already outstripped its carrying capacity will lead to catastrophe. We all discount the future. But temporal myopia cannot undo the certainty that our most successful, personal pursuits will come to an abrupt end in three or four score years at best,. After that, the only impact we can have is a legacy. Today the prospect confronts us that our chief legacy may be a world in ruin.

Scientific research has confirmed what history has consistently taught: average people can be the authors of great good or great evil. It largely depends on the circumstances in which they find themselves. As a species, we have the opportunity and responsibility to create the *oikos* within which we can flourish, an economic habitat where we find it easier, and more fruitful, to do good by doing well.

5 Politics

An embodied being can never relinquish actions completely; to relinquish the results of actions is all that can be required.
—Bhagavad Gita

Sophocles's *Oedipus Tyrannus* tells the famous tale of a king pledged to rid his kingdom of pollution. But King Oedipus is no environmentalist. The pollution in question is a person, the suspected source of a plague that wracks the city of Thebes. Years earlier, the former king of Thebes was murdered, and the culprit was never caught. This murderer at large constitutes a form of religious pollution that must be extinguished for Thebes to regain its health and stature. In his ruthless pursuit of the culprit, Oedipus learns the most terrible truth imaginable.

The myth that inspired Sophocles's classic tale was known to every Greek. Oedipus was orphaned at birth by his parents, the king and queen of Thebes, owing to a prophecy that the young prince would kill his father. The baby, exposed to the elements and left to perish, is secretly taken in by strangers and reared in a foreign land. As a young man, Oedipus learns of the prophecy. But he falsely believes his adoptive parents to be his true mother and father. To avoid the grisly fate slated for him, Oedipus leaves his assumed kin and quits his adopted land. But his valiant efforts to construct a new ending for the forecasted script are in vain.

Meeting a man at a crossroads who will not yield the way, the young Oedipus—prideful and quick to anger—slays the stranger. The unrecognized traveler, of course, was Oedipus's father, King Laius of Thebes. Unaware that he has become a parricide, Oedipus continues on his journey. Quick-witted and ambitious, he tackles a dangerous challenge no one before him has met: solving the famed riddle of the Sphinx. This heroic act frees Thebes from the grip of the mythical monster. Oedipus enters his home town a hero and takes the recently widowed queen as his bride.

A parricide married to his mother is a most unnatural and unholy state of affairs, and it brings a plague on Thebes. Against all counsel and ignorant of what he is doing, Oedipus vows to rid Thebes of the polluter. In his determined effort to root out the cause of the plague, King Oedipus eventually discovers his own identity, as does Queen Jocasta. Following the suicide of the horror-stricken queen, his mother and wife, Oedipus pierces his own eyes and flees the land. The once mighty ruler is chastened and brought low, a blind man led away to a meager life in exile.

How are we to understand the moral of this tragic tale? To modern ears, it seems wholly unjust that an innocent child would be shackled to such a gruesome destiny. Equally unjust is fate's rebuff to each of Oedipus's dogged efforts to shield himself and his kin from calamity. Try as he might to avoid the horrors of the foretold tale, every heroic act brings the cunning and strong-willed man closer to the climax of the macabre plot.

Sophocles was iterating a common theme of the Greek poets: hubris, or overweening pride, brings the wrath of the gods. The Greeks loved their heroes, who often exhibited hubris, so the poets took on the task of exploring the thin line that separates admirable talent, dedication, and ambition from dangerous pride and stubborn willfulness. The Greek gods were a jealous lot who did not like to see mere mortals assume the prerogatives of deities. Human beings were not meant to be self-sufficient. Theirs was a life of interdependence. To be born a human is not to control one's parentage. To be a human child is not to control one's upbringing. To live in a human community is not to control one's destiny. Men who hubristically sought a god-like freedom from these webs of relationships would discover—often too late—that such mastery was not their appointed lot. Eventually wisdom would be gained. But as the ancient playwrights reminded us, that wisdom was often tardy and gained at the price of great suffering.

A brilliant mind and lusty spirit leaves the young Oedipus convinced that everything is within his grasp. Not being a god, however, he can never do just one thing. He tries in vain to recast the gruesome role that he is destined to play. But as the famous Greek tragedy so eloquently describes, every attempt Oedipus makes to avoid his fate by willful action causes him to fall more tightly into its grip. Oedipus does not realize that as a mortal, his every action will have unintended consequences. By killing a stranger, he becomes a parricide. By marrying a queen, he engages in incest. By relying solely on his own wits, he is prevented from acting wisely. By exercising tyrannical sovereignty, he dooms his reign. By attempting to master his fate, he becomes its victim. The side effects of Oedipus's purposeful actions cause him to fulfill a destiny he rightly abhors as most terrible. In the end, Oedipus must blind himself that he may truly see and banish himself that he might seek community.

The last character to speak in Sophocles's tragedy, Creon, counsels Oedipus not to pursue mastery in everything he does. These words of wisdom are tardy indeed. Oedipus tried to control his own life's tale. He attempted not only to script his present and future, but, effectively, to rewrite his past. The pursuit of sovereign control over worldly life by mere mortals is not tolerated by the Olympian gods. Such hubris ensures their wrath. Tragedy is inevitable.

The *Zoon Politikon*

If we were all-powerful, self-sufficient gods rather than mortal men and women, we would not need politics. Alternatively, if we were spiders hatched from eggs and equipped to survive on webs we alone build and oversee, we would not need politics. But we are neither self-sufficient gods nor animals whose instincts fully equip them for life. Instead, we are creatures who must rely from birth through old age on the care and cooperation of others to survive. Indeed, we rely on our communities to become full human beings. The traits that we deem definitive of our very humanity—the power of speech, reason, moral values, technological skill—can be gained only by way of our partaking of community. None of these capacities could develop, and we could not benefit from their exercise outside the social networks that make us who we are.

Aristotle defined a human being as a *zoon politikon*, a political animal. Human beings, the Greek philosopher observed, form political communities in order to survive. Once formed, however, these communities do not simply satisfy our most basic needs. They make possible not only bare life, but the pursuit of the good life. In satisfying their most basic needs for food, shelter, and security, and even more so in gaining the benefits of culture, human beings find themselves embedded within vast networks of social relationships. These networks are the *oikos*, the habitat, of the *zoon politikon*. In this respect, an investigation of our political nature provides great insight to the fundamental interdependence of our lives.

A biological community, or ecosystem, is defined as a set of interdependent organisms. Likewise, a political community or polity is a group of interdependent individuals. As political animals, we live within and participate in shaping a connected world. Politics is what we employ to organize our common lives. This organization is achieved by marshaling consent, stimulating cooperation, regulating competition, and publicly validating necessary coercion. The word *politics* derives from *politica*, which refers to the *polis*, the ancient Greek city-state. When the Romans translated *polis*, they settled on the term *res publica*, or republic—literally the "public thing" or "public affair." Politics concerns the pursuit of public goods.

Were we identical clones following instinctual drives, much like a hive of bees or a nest of ants, we would not need politics to regulate our collective activities. Instinct would suffice. Human communities, however, are composed of unique individuals who require much training and education to survive and prosper. For all our distinct traits and matchless personalities, we develop into recognizable persons owing to our involvement in social environments. It is in this regard that Aristotle would insist in *The Politics* that "the whole is of necessity prior to the part."[1] The part, the individual, becomes who he or she is by way of the whole, the political community. The human organism becomes a distinct individual only within a web of social relationships. Human beings require community to become unique individuals. Shaping this community and ordering the interactions of the unique individuals composing it is the stuff of politics.

To survive and prosper, communities must find a way to organize themselves, provide and distribute goods, regulate interactions, maintain security, and safeguard individual rights and liberties. Politics refers to the means by which individuals or groups of individuals (such as political parties) gain, maintain, exercise, and distribute power to serve these purposes. Politics is always and primarily about power. Power is political in nature when it serves the procedures and institutions by which binding decisions are made for society. These procedures and institutions are known as government.

Aristotle considered the study of politics to be the master science. He did not bestow this accolade on politics because its study allowed for accurate prediction and great certainty. Quite the opposite was true. Aristotle insists that politics, pertaining as it does to the interactions of diverse, unique individuals, will not allow accurate forecasts or universal statements. Politics is a dynamic affair in which diverse individuals engage in inimitable interactions. Context is crucial and change inevitable. Given these constraints, practical wisdom is the best means of navigating the political realm. But if politics does not allow prediction and certainty, why would its study constitute the master science?

The study of politics earned Aristotle's esteem owing to its scope. Politics orders the social realm as a whole, and in doing so allots to every other human activity and relationship its appropriate functions and limits. In this respect, the other sciences depend on political science to determine their rightful place within the human *oikos*. Other fields of study and facets of life also rely on politics to set their boundaries. This is not to say that politics controls, or has the prerogative to intrude on, every aspect of our lives. But that is only because politics also sets limits for itself.

Politics (and explicitly democratic politics in a constitutional regime) sets itself clear limits. It does not determine, for example, what citizens should believe regarding the existence of God or the fate of their souls. Rather, it is the process whereby citizens establish and secure the freedom to determine their own religious beliefs. Likewise, through the creation of legislative policies, politics establishes and safeguards the diverse realms of primary education, standards of health and welfare, and the infrastructure that allows and fosters economic opportunity. If the rule of law and personal and civil liberties are to be maintained, then politics—through mechanisms such as constitutions and bills of rights—achieves this feat.

In this respect, politics is an indispensable feature of our lives even when we are most concerned to preserve our privacy and maintain our independence from (partisan) political influence. If we want to safeguard a realm of privacy in which politics cannot intrude, then we must act politically to achieve this good. As political theorist Bernard Crick observes, "The price of liberty is more than eternal vigilance . . . it is eternal commitment to political activity."[2] At the end of the day, our right to avoid or circumscribe politics must be politically secured. There is no alternative. Politics matters, and should matter, even to those who are most inclined to remain politically uninvolved and inactive.

We investigate politics in this chapter for two reasons. First, public goods cannot be secured without politics. A healthy environment is one such public good. Economic opportunity and security, social equity and empowerment, and cultural creativity and learning are other public goods. In this respect, sustainability—the balanced pursuit of these goods—is inherently a political endeavor. Second, exploring the nature of political life can help us understand how living in a connected world rife with unintended consequences does not have to impinge on our freedom. In fact, it is the only way our essential freedom can be experienced.

Democratic Politics, Peace, and Sustainability

To say that politics matters is not meant to underline the importance of partisan differences in our political lives, though these differences and the struggles they give rise to may be important. Rather, it is to observe that politics, in the broadest sense, shapes our very humanity. More specifically defined—in terms of the form of government and character of a citizenry— politics is also a matter of grave significance. Whether a political regime is democratic, and how democratic it is, matters. Certainly it matters to citizens, who enjoy greater liberties, rights (including the right to privacy), and security in democratic nations than in nondemocratic regimes. The fact is that democracies are less prone to the arbitrary arrest of citizens and kill fewer of their own citizens than autocratic governments do. Given that the government-sponsored killing of citizens causes six times as many deaths in the world as does war between states, the civil protections offered by democratic constitutions are significant indeed.[3]

Beyond establishing more peaceful relations between rulers and ruled, democracies also prove to be more peaceful in their relations with other states. At least, democratic nations tend not to wage war against each other. War is destructive of human lives, of human communities, and of the web of life on which individuals and communities depend. If democracy reduces the incidence of war, then the study and practice of democratic politics is inherently related to sustainability.

In 1795, Immanuel Kant penned an essay entitled "Perpetual Peace." Here he speculated that constitutional republics were unlikely to war against each other because the majority of their citizens, likely to suffer the consequences of hostilities, would only support going to war in self-defense. Beginning in the 1970s, social scientists began testing this thesis. They aggregated tremendous amounts of data on interstate violence and analyzed the results.[4] Their analyses have been largely supportive of the claim that democratic nations do not fight wars against each other.

There are problems with the democratic peace theory. Uncertainties arise because many forms of conflict (e.g., quick or small skirmishes, non-lethal military stand-offs, isolated exchanges of firepower) are sometimes excluded from consideration. Ambiguities and controversies also arise because some countries may be excluded from the analysis based on the criteria for being a democracy. Democracy is a continuum. By any criterion we might choose, there are more democratic and less democratic regimes. In turn, no universally accepted criteria exist to determine what minimally is, or is not, a democratic regime. The most widely accepted basic qualification is a representative government wherein citizens elect their political rulers under the protection of a binding constitution. But there is significant scope for disagreement as to how frequent, fair, and free elections must be, and how much of the adult population must effectively and actually be able to vote. There is more disagreement about what other civil and economic rights citizens should hold under constitutional protection, and how effective the legislative branch of government must be at controlling the executive branch to constitute a truly democratic regime.

Nonetheless, the data confirm that there has never been a war between well-established democracies. Moreover, democracies of all sorts are unlikely to fight each other. This is not to say that democratic regimes do not fight wars. Since the mid-twentieth century, some of the most war-prone countries—the United States, Israel, and the United Kingdom—have been democracies. In turn, newly democratizing countries appear to be even more prone to war than stable autocracies. But however prone they are to violent conflict with nondemocratic states, democratic regimes are unlikely to take up arms against each other.

For all its qualifications, the democratic peace theory may provide the best example of an empirical law—a statement that provides an accurate description of current reality and is capable of predicting future events—in the field of international relations.[5] That may say more about the inability of this field of study to develop law-like generalizations than it underlines the value of the democratic peace thesis. To be sure, relevant data are sparse until you get to the twentieth century. History has provided few examples of democratic countries and even fewer examples of mature and stable democratic regimes. So although the data are encouraging, only time will allow us to say with greater certainty whether democratic nations wage war against each other. In the meantime, it would be dangerous indeed if the democratic peace thesis were employed to justify aggressive wars against autocratic regimes in the belief that making the world "safe for democracy," in Woodrow Wilson's famous phrase, would ensure lasting peace.[6] At the same time, it would be wise to welcome the birth of democratic regimes abroad and, as important, strengthen democracy at home in the interests of a peaceful world.

The relationship between democracy and sustainability is at least as uncertain and controversial as that between democracy and peace. Democratic forms of deliberation and interaction promote a forward-looking and expansive outlook.[7] To be a democratic citizen is to accept responsibility to participate with a diverse constituency in deciding the future of your polity. Such an inclusive, future focus would appear well aligned with the values of sustainability. But democratic theory does not always correspond well to democratic practice. And with regard to the consumption of natural resources and the production of greenhouse gases and other wastes, democracies, with the United States at the forefront, have been the world's worst offenders. In part, this nasty reputation may simply be a product of the fact that democracies over the past half-century have been the wealthiest countries in the world. Their strong, growing economies, grounded in the rapid exploitation of natural resources and fossil fuels, have been exemplars of overconsumption and the unsustainable production of waste. That is not to excuse democracies for being unsustainable. But it may say more about what human beings are prone to do when given the chance than what democratic practices and governments themselves promote.

At the same time, the concept of democracy today is tightly coupled with the expectation of ever-rising levels of consumption, in part because democratic leaders frequently promote economic growth as a panacea. Adding to the problem, democratic regimes can be every bit as nationalistic as autocratic regimes. Such nationalism can undermine efforts to promote global environmental stewardship. Finally, democracies safeguard their citizens' economic freedoms, and these safeguards are often abused by individuals and corporations in an unsustainable pursuit of profit. To the extent that democracy fosters consumer culture, growth-oriented economies, nationalist sentiments, and unregulated greed, it may thwart efforts to develop more sustainable societies within a global web of life.

Notwithstanding these tendencies, democratic regimes have established the world's most advanced, far-reaching environmental laws and stand at the forefront of environmental protection both domestically and globally. In turn, many of the worst cases of environmental degradation have occurred in autocratic regimes, such as the former Soviet Union. Empirical analyses provide qualified support for the thesis that democracies pursue and foster environmental sustainability better than autocracies do.[8] In large part, this is because the vibrant social movements and nongovernmental organizations of civil society that demand environmental caretaking in democratic regimes are much less developed and active in autocratic regimes.

Environmental protection is a central feature of sustainability. However, sustainability is grounded not only on ecological health but also on social justice, that is, the equitable distribution of social goods. To be sustainable, a social system must foster an equitable sharing of resources and power. Without an equitable distribution of power, it is unlikely that an equitable distribution of other social goods could be achieved or would be maintained.

As an empirical fact, and, one might say, by definition, democracies outperform autocratic regimes in terms of the equitable distribution of political power. In turn, certain democracies perform better at securing other forms of social justice. How equitably resources and goods are shared in a polity largely depends on how equitably political power is shared. In the absence of a democratic sharing of power, the least advantaged in society are typically further disadvantaged in terms of the distribution of goods by the most powerful individuals and groups who can pursue their self-interest without oversight or regulation.

Democratic government in theory is transparent. Its decision-making processes are available to examination by the general public. It is open government, characterized by parliamentary debate held in open (as opposed to closed-door) sessions, open legislative (roll call) votes where the public knows who voted for or against each bill, and open records, including access to lobbying and campaign finance information, so the public knows how elected officials may have been influenced in their decision making even before they entered legislative chambers.

Transparency is also a crucial feature of sustainability. The only way to ensure that practices are sustainable is to submit them to critical scrutiny. Full-cost pricing of goods and services, for instance, is possible only when social and environmental impacts are made visible. The transparency that provides a foundation and safeguard of democratic government meshes with the transparency that is required to pursue more sustainable livelihoods and economies. In both cases, transparency allows people to be more widely and meaningfully involved in the process of deliberating and securing the common good. The seventeenth-century polymath Francis Bacon famously observed that "knowledge is power." Transparency in politics and in the pursuit of sustainability empowers through information.

There is a final, and perhaps even more salient, parallel between democratic politics and sustainability. Both are grounded in the recognition and embrace of interdependence. That sustainability is grounded in interdependence has already been established. That politics, and democratic politics specifically, is grounded in interdependence is the thesis we now explore. Political interdependence is primarily manifested in the exercise of power.

The Nature of Power

Power is often understood as a form of control. As such, it signals a power over other individuals or groups. One exercises power by getting people to do, say, believe, or value something that they otherwise would not have done, said, believed, or valued. The arts of persuasion are key to such influence, and statespeople are duly recognized as powerful in this respect. They are adept at rhetoric, parliamentary debate, and speechmaking. When statespeople prove sufficiently persuasive, their arguments and recommendations become transformed into policies. One might think of governmental policies as petrified forms of persuasion. They are formal, written prescriptions for practices, much like social custom is an informal, unwritten prescription for practice.

Sometimes the power of persuasion is not enough. Sometimes, it appears, the only way to get people to do certain things (or not do certain things) is by threat of force. Niccolò Machiavelli held that good politics ensues from good laws. But you cannot have good laws, Machiavelli insisted, without good arms to back them up. Behind every political force, lay the force of arms. Likewise, Thomas Hobbes, the seventeenth-century British political theorist, argued that covenants without swords were but words, with no capacity to secure themselves. Laws, it follows, both dictate what people are to do or not to do and spell out what coercive force will be applied if the law is not followed: fines or imprisonment or, in certain cases and regimes, capital punishment. For this reason, coercive force or violence is often viewed as the ultimate form of power.

National governments, established and maintained by political means, claim to hold a legitimate monopoly on the use of force within their borders. Governments employ their coercive power through judicial systems, the police, and armed forces to maintain order, secure public safety, try offenders, and incarcerate lawbreakers. Coercion and violence can make people do things they would not otherwise do. The threat or use of force alters behavior. However, politics is not fundamentally a phenomenon of coercion or violence. Indeed, violence marks the boundary of the political. A tyranny, in this sense, is not a political regime. The tyrant, who rules by force alone, is not a political leader. Politics is fundamentally constituted by speech and action, that is, by persuasion, coordinated deeds, and lawmaking. The use of coercive force in society typically gains legitimacy through political means. But brute force is not politics.

Brute force is a one-way street. Politics, in contrast, is a relationship of coinfluence. Whenever power is in play, there is generally an imbalance of influence. Political power is seldom, if ever, wholly egalitarian; someone typically has the upper hand. But the person being influenced is able to respond or resist. A political relationship often has the threat of force in the background. But it ceases to be political once violence takes away the option of response or resistance.

Having the upper hand can be alluring, and corrupting. As the nineteenth-century English historian and moralist Lord Acton famously observed, "All power tends to corrupt, and absolute power corrupts absolutely." With this in mind, the American religious and social thinker Reinhold Niebuhr argued for the sharing of power through democratic institutions. "Man's capacity for justice makes democracy possible," Niebuhr wrote, "but man's inclination to injustice makes democracy necessary."[9] Whenever power is exercised, some people's interests are typically better served than those of others. The exercise of power is not always benign, and it often is wielded for self-serving purposes. It follows that a political system that distributes and balances power is better prepared to serve justice. Representative democracies distribute and balance power between elected officials of varying viewpoints and parties, between elected officials and the citizens who decide their electoral fates, and, in democratic republics, among the three branches of government (legislative, executive, and judicial).

Political power is often wielded to serve particular interests through the policies, institutions, and laws it creates and maintains. Equality under the law is a foundation of liberal-democratic regimes. But like most other, if not all, products of power, the laws of the land—even when universally and fairly enforced—do not equally benefit all. Anatole France, the nineteenth-century French novelist, observed that "the law, in its majestic equality, forbids the rich as well as the poor to sleep under bridges, to beg in the streets, and to steal bread." The legal edict that forbids such desperate acts protects well-to-do citizens from the disconcerting sight of homeless people, the annoyance of panhandlers, and the aggravation of petty theft. This law provides fewer benefits for the impoverished. The ancient Romans insisted that the key question to ask of any political activity was *Cui bono?*(Who benefits?). To say that politics concerns how we define and order our collective lives does not mean that all have equal voice or that the results will serve all equally well.

Political scientist Harold Lasswell's classic book, *Politics: Who Gets What, When, and How,* asserts that political power is wielded in a competitive struggle between individuals or groups to secure particular needs and wants.[10] There will always be winners and losers in this game of persuasion, bargaining, logrolling, manipulation, strategic maneuver, and rule making. Given a world of scarce resources, however, politics provides a crucial means of avoiding violent conflict over the distribution of goods. Political life is the alternative to what Thomas Hobbes called a "war of all against all" that would condemn us to incessant violence and, ultimately, to an existence that is "nasty, brutish, and short."[11]

Politics allows us to determine who gets what, when, and how through means other than brute force. However, politics is not restricted to determining how to divvy up the pie. It also plays a key role in identifying and gathering the ingredients for the pie, and getting it collectively made and baked. As important, politics plays a key role in determining how our needs and wants as pie makers and pie eaters are perceived and defined, how these needs and wants get generated and transformed, and how they are pursued. Politics goes well beyond the simple allocation of scarce resources because it affects how we come to define these resources and understand ourselves as individuals and members of a community. Politics articulates, secures, and transforms not only interests but also identities.

With the onset of political thought in ancient Greece, it was clear that politics meant more than mitigating conflict and fostering cooperation in the distribution of resources. Plato maintained that politics was chiefly "concerned with the soul."[12] Statecraft, for the ancient Greeks, was a form of soulcraft. Legislators involved in the creation and maintenance of the city-state were effectively tasked with shaping the character of the citizenry. For the Greeks, this ordering of souls constituted an education in virtue. Both Plato and Aristotle agreed that such an education was the first task and greatest responsibility of any legislator.

Uneducated citizens cannot well organize their collective life. And an ill-organized polity, one that has bad laws, will damage the character of its citizenry. So while insightful soulcraft is an essential part of statecraft, skilled statecraft is what allows the practice of soulcraft. The structure of the political regime molds the characters of its citizens, and the caliber of the citizenry determines what sort of polity can be formed and maintained. In the interdependent realm of statecraft and soulcraft, you can never do just one thing. Each policy and each piece of legislation not only reorganizes the political regime, it also affects the development of its citizenry. In turn, every successful effort to educate citizens has the effect of transforming the polity.

In its dual role of allocating scarce resources without violence and shaping the beliefs, values, identities, and relationships of the actors involved, political power involves the exercise of influence. It generally serves some interests more than others and hence constitutes a form of "power over." But the exercise of power always establishes a relationship. The mutual influence manifested in such relationships constitutes a form of "power with."

Power with, as the term suggests, is not an individual achievement. It pertains to our capacity to act with others to achieve what could not or would not have been achieved alone. Cooperation is a patent form of power with. When we cooperate, we come together to attain some specific end that would have proven difficult, if not impossible, to attain as isolated individuals. Unlike power over, the exercise of power with signals that the actors are not playing in a zero-sum game: one person's gain is not, by necessity, another person's loss. Rather, cooperation produces a positive-sum game. Politics, as an exercise of power with, underlines our mutual needs and our common means of satisfying them. It responds to, manifests, and strengthens our interdependence.

Cooperation, and the power that it generates, requires structure. It will develop, or at least fully develop, only when the game being played is organized so that mutual benefits are possible. In a zero-sum game, such as a fight to the death, cooperation will not develop. But even in positive-sum games, cooperation may fail to develop if the problems inherent in collective action are too pronounced. Whenever people work together, there is a temptation for some to free-ride on the contributions of others. If it is possible to receive all or most of the benefits of a collective project while contributing nothing or little, free riders will likely emerge to take advantage of the system. The opportunity for free riding is a collective action problem. Structuring a system to mitigate free riding strengthens cooperation.

Consider two people who need to get across a lake. There is no shortage of boats, but a strong headwind makes it impossible for a single rower to make progress across the water, so the travelers decide to cooperate. They get in a large rowboat, each grabs oars, the one behind the other facing backward, and they proceed to row. While the person at the front of the boat must row for the boat to make progress across the lake, the level of his effort cannot be perceived by his traveling companion, who has her back toward him. So the rower at the bow may be tempted not to dip his oars very deeply in the water or pull with all his strength, as the craft will still get to the other side without his full exertion. The organizational structure chosen for the cooperative effort, a form of sculling, is susceptible to free riding.

Alternatively the boaters might choose to sit side by side, each taking one oar with two hands. In this case, they would be sweep rowing rather than sculling to the other side of the lake. Equal exertion is required for sweep rowing; if one partner slacks off, the boat will move in large circles.

Whether sculling or sweep rowing, the boaters find themselves in a relationship of interdependence. To enhance cooperation, however, relationships of interdependence need to be structured to mitigate collective action problems such as free riding. They must be as equitable as possible. Getting the structure of the system right—organizing well the rules of competition, the avenues for cooperation, and the system of costs, benefits, rewards, and penalties—is crucial. Getting the structure of the political *oikos* right allows a productive balance in the exercise of power with and power over. This is where grounding education, policy, and law in ecosophic awareness can transform a democratic political regime into a sustainable democracy.

Education, policies, and laws clearly affect the organization of relationships within a political regime. As we saw in the previous chapter, tax policies, laws concerning transparency such as right-to-know legislation, and educating citizens in their roles as residents and consumers can have significant impacts on how sustainability is pursued. Perhaps as important, a political education—that is, a deeper understanding of the nature of politics and power—can help citizens appreciate how freedom, the most cherished political value, is best realized by taking action within webs of interdependence.

Action and Freedom

Hannah Arendt insightfully explored the interdependencies of political life. Arendt was a Jewish intellectual who fled Nazi Germany to Paris in 1933, and subsequently fled occupied France to the United States in 1940. She became one of America's most distinguished political theorists. Arendt deems action a unique human capacity and the sine qua non of political life. Taking action allows us to experience freedom. Political life, first and foremost, is a life of action.

To act, for Arendt, is not simply to do something. Painting a portrait, designing a building, or constructing a bridge is not acting. These are forms of work that produce useful and perhaps beautiful artifacts, but they are not action. Likewise, to labor is not to act. Earning one's living by driving a taxi, stocking store shelves, or picking up trash is toil, but it is not action. Unlike labor or work, which can be and often is carried out by individuals in isolation, action necessarily occurs among and between people. For

Arendt, the condition of political life is plurality. The political *oikos* is inhabited by diverse (groups of) individuals. Action occurs within this plural world not by dint of the aggregated efforts of isolated individuals but through the "*sharing* of words and deeds."[13]

Action is the essence of power with. It highlights the interdependence that occupies the core of political life. This interdependence arises in part because people come together to achieve ends they otherwise could not achieve as individuals. But the essence of action is not the attaining of fixed goals, whether these goals are achieved by individuals or are the product of cooperative efforts. Indeed, Arendt insists that action truly occurs only when the instrumental pursuit of identifiable ends ceases. That is a puzzling statement. Just as puzzling, Arendt maintains that human freedom—the product of action and the most cherished political value—is made possible only by way of interdependence. These puzzles need to be solved together.[14]

In an essay entitled "What Is Freedom," Arendt writes that "the appearance of freedom . . . coincides with the performing act. Men *are* free—as distinguished from their possessing the gift for freedom—as long as they act, neither before nor after; for to *be* free and to act are the same."[15] In developing this understanding of freedom, Arendt challenges traditional notions that identify freedom with an individual who remains "master of his acts."[16] Freedom is often understood as an ability to do what one sets out to do, unhindered by external obstacles. From this viewpoint, freedom is a form of sovereignty. We are free when we become masters of our fate, beholden to no foreign powers and subject to no external constraints. Freedom is the ability to act with independence in the service of one's interests and preconceived ends. Arendt rejects this notion of freedom. Far from being its prerequisite, she insists, sovereignty signals the demise of freedom.

Sovereignty denotes control of outcomes. It is the capacity to set one's own course and, ultimately, to control the effects of one's actions. Arendt denies that freedom is manifested in the internal will to attain particular ends or in the absence of obstacles to the achievement of these goals. She writes: "Under human conditions, which are determined by the fact that not man but men live on the earth, freedom and sovereignty are so little identical that they cannot even exist simultaneously. Where men wish to be sovereign, as individuals or as organized groups, they must submit to the oppression of the will, be this the individual will with which I force myself, or the `general will' of an organized group. If men wish to be free, it is precisely sovereignty they must renounce."[17] Rather than a product of a sovereign will, Arendt argues, freedom is a public event that escapes individual or collective control. It appears whenever citizens act together and generate novel aspirations, relationships, and events in the wake of their interaction. When people act in concert, they experience freedom in the very originality and unpredictability of the results. Action instigates freedom by allowing the new and the unforeseen to occur. To act, and to experience freedom, is not to control the effects of one's actions but to initiate unintended consequences.

We will take a step back and examine why freedom is not, most fundamentally, a product of our ability to fulfill desires or exercise will in the absence of obstacles or constraints. First, we must ask how we came to have a particular desire or settle on a particular object of will. Everything—including desire and willful intention—has a cause. Desires and intentions may be stimulated by the sight, sound, or smell of an object, or perhaps from its very idea. One sees the ripe red plum, a desire to eat it wells up, and your hand reaches out to pluck it from the tree or table. Or one gets the idea of playing catch and, after some consideration, sets oneself to the pleasant task at hand. Neuroscientists can lead us through the chain of physiochemical activities in the brain that produce a sense of desire or a willful intention. These neural activities are initiated from perceptions or thoughts and result in an object of desire firmly in hand or in pursuit. But in what sense can a chain of physiochemical reactions in our brains be said to manifest freedom? Do we believe that our enjoying the pleasures of eating, sleeping, and recreation demonstrates our freedom when the same behavior exhibited by one's pet beagle—which clearly also enjoys eating and playing catch—does not? And what of the behavior of ants, bees, and spiders or, for that matter, bacteria and viruses? At the cellular and molecular level, every action is the product of a chain of previous physical interactions. How and where can freedom be located within this chain of events?

The problem with positing freedom as the ability to satisfy one's desires or carry out one's will in the absence of constraints is that it does not account for the actual formation of desire or will. And this latter event, when subject to scrutiny, will always lead one down an extended chain of other events. Pushed far enough, it will take us to the level of the physio-chemical reactions of cells and molecules. Of course, one can simply insist that humans have a free will that beagles, spiders, and viruses do not, regardless of our inability to locate or adequately define it. But that is rather unhelpful. It simply begs the question of freedom.

Arendt argues that human freedom cannot depend on our ability to escape from a chain of events and arrive at an uncaused cause.[18] The only candidate for such a wholesale break from cause-and-effect relationships would be a random event, something that escapes all causal chains by way of its spontaneous, chaotic origin. But how satisfying is it to posit our freedom as the product of the random firings of brain cells or, at a deeper level, the random collisions of molecules and atoms?

Arendt suggests that we conceive freedom differently. Freedom does not arise because we break free of causal chains and effectively escape the past. But it does constitute a new beginning. Arendt asks us to conceive freedom not as a renunciation of the known and settled past but as an invitation to the unknown future. Freedom arises whenever actors "begin something new and . . . [are] not . . . able to control or even foretell its consequences."[19] Politically speaking, one acts freely when one initiates events without determining their outcome. The capacity to begin something new without gaining sovereign control over it, Arendt claims, is "the supreme capacity of man; politically, it is identical with man's freedom."[20]

With this understanding of freedom in mind, Arendt contrasts politics to administration. When we administer or manage something, we attempt to control the results of action. Clear goals are set, and effective means to achieve them are devised. Success depends on crafting and carrying out plans that efficiently attain predetermined ends. The "administration of things" pertains to all affairs where such means-ends relationships can be ascertained with certainty. It is a technical task. Arendt contrasts such administrative behavior to the political realm of "things which . . . we cannot figure out with certainty."[21] Politics is more like conversation than administration. When we truly converse with another person, we are not simply intent on winning an argument by marshaling facts and using rhetorical skills. In a true conversation, we do not know where our interaction will take us. Rather, the verbal voyage sets its own path. A conversation might exhibit various twists and turns, and, as a product of much back and forth exchange, it may produce any number of agreements, disagreements, revelations, resolutions, queries, or conclusions. But it has no predetermined destination. Likewise, politics is not the management of collective life in pursuit of preconceived ends. Rather, it is an unpredictable interaction of unique individuals. These individuals discover freedom in the very unpredictability of their interactions.

When we exercise sovereignty or mastery, we seek freedom from uncertainty. We instrumentally engage in efforts designed to attain given ends. We want to know that our desires will be satisfied and our intended goals reached. And we do not want to be interfered with in pursuing these desires and goals. Political life is not an exercise in sovereignty. It does not achieve freedom from uncertainty. Rather, it manifests freedom for uncertainty. As dwellers of a political *oikos,* our actions serve as open invitations to co-inhabitants. In acting we are inviting others to respond.[22] How they will respond, and how we will in turn respond to their responses, remains indeterminate. It is precisely this uncertainty, this unpredictability borne of the diversity and uniqueness of human actors, that allows the experience of freedom in political life.

The novelist Ernest Hemingway once said that if he knew the answer, he would not have to tell the story. What Hemingway says about the fictional novel applies equally to political action. "If we can figure it out with certainty," Arendt rhetorically asks about political life, "why do we all need to get together?"[23] To act politically is to act in concert. This interactive effort invites the unknown. As actors, we initiate rather than administrate.

Political action is not simply what an actor says it is or intends it to be. The meaning of an action is not to be found in individual intentions or aspirations. Rather, the meaning of action is found in the responses it provokes. It is a product of what action does in the world. When the *zoon politikon* acts, she enters a web of relationships and sends ripples through it. There are responses to her action and responses to these responses. Effects and side effects abound. And while the initiator of action might want to privilege the effects she intended to produce, the side effects may be as, if not more, important. Just as squirrels hiding acorns are not doing just one thing, neither is the political actor. And from a political point of view, as from a systems perspective, it makes little sense to privilege intended effects from side effects. Indeed, to the extent that we are interested in the experience of freedom, it is the unintended consequences that are most salient.

In the end, the meaning of an action is less determined by its author's intentions than by the host of responses it stimulates. When acting in the political realm, we are initiating that which cannot be predetermined. We are partaking in what natural scientists call *emergent behavior* (which I address extensively in chapter 7). Emergent behavior occurs when the activity of a whole is more than the sum of the activities of its parts. Something new, different, and unforeseen occurs—something initiated by a specific actor or actors but neither controlled nor concluded by them.

Action is not the whole of politics. Rather, it is the leaven that gives life to a recipe whose binding agents are habits, customs, conventions, rule following, and the various behaviors and reactions they stimulate. Distinguishing "free" action from behavior, Arendt observed that "most of our acts are taken care of by habits, just as many of our everyday judgments are taken care of by prejudices."[24] When we do something habitually, we are reacting to a stimulus rather than acting freely within a context. The lion's share of our lives is defined by such habits and prejudices. These routine forms of behavior and thought allow us to become very skilled and proficient in what we do. Brushing our teeth, riding a bicycle, building houses, calculating sums in one's head, and crafting plans are all good habits and skills to develop. But they are not forms of action.

A political action, in this respect, is inherently different from making an artifact or carrying out any other instrumental effort. When crafting an artifact or implementing a plan, we are making something according to a preconceived idea, model, or blueprint. We know ahead of time what we are trying to achieve, and our success depends on most closely approximating our model or blueprint. Well-honed skills, acquired through habit, allow proficiency in our efforts. In this respect, the meaning of our technological endeavors remains largely within our control (though, as we have seen, how technological artifacts subsequently get used and misused is not). In contrast, when we act in the political world, we are not following a plan. We are soliciting responses. To be sure, political actors have specific intentions, hopes, and goals. But to the extent that we act politically, we enter a network of interdependent actors and invite a conversation of deeds. The meaning of these interactive events can be determined only retrospectively, when a story is told about their impact on the world.

Politics and Narrative

Hemingway's remark about not needing to tell the story if he knew the answer applies to political life in more ways than one. Political action might be best understood as the beginning of a story. Action is the onset of plot. More accurately, action is the start of a coauthored story without knowable ends. It begins a story on its way and invites responses. Political life is not a plan or pattern to be realized through efficient administration. It is, in large part, an open-ended narrative. Action, the essence of political life, constitutes the beginning of a tale whose meaning emerges only by way of the contributions of others. The meaning of an action, in other words, is found in the story we tell about the responses it evoked.

Consider Martin Luther King Jr.'s "I Have a Dream" speech, delivered from the steps of the Lincoln Memorial on August 28, 1963, during the March on Washington for Jobs and Freedom. The meaning of King's speech cannot be found in the mere description of when and where it was delivered. Neither do the words of the speech nor a biographical account of its author exhaust its meaning. Rather, the meaning of King's powerful speech is found in what it did in the world: how it reflected and stimulated the aspirations of a nation mired in oppression and injustice to rise above racial prejudice. The meaning of King's speech, in other words, is to be found in the narratives that capture its personal and political context, its historical significance, and its concrete impact.

The word *narrative* derives from the Latin *narrare*, which means to relate, and the Greek *gno*, which refers to knowledge. To enter a narrative is to know relationships. Narratives display the relationships of parts to other parts and the relationships of parts to the whole. Stories allow us to discover meaning because they reveal action in terms of its antecedents, its effects, and its side effects. As Arendt insists, "Action reveals itself fully only to the storyteller."[25] Before they are put in a narrative framework, actions would simply be what Arendt calls "sheer happenings."[26] Without a story to lend them significance, they would appear as an incoherent and incomprehensible whirl of activity. Translating sheer happenings into meaningful actions requires placing them in a story with beginnings and endings, protagonists and antagonists, plots and themes, conflicts and resolutions.

Once we accept that the meaning of action is not controlled by its author but is found in the story that chronicles and celebrates it, Arendt asks us to take another step. Having acknowledged that the actor cannot control the meaning of her action, we can infer that the actor also cannot control the meaning of herself. Like Oedipus, the political actor reveals herself through her actions. But this self-revelation is not a sovereign act. Arendt insists that "nobody knows whom he reveals when he discloses himself in deed and word."[27] An actor's (self) disclosure provokes unpredictable responses. The meaning of the events she initiates no less than the ultimate identity of the actor are determined only in retrospect, once responses have been aired and the reverberating web of human relationships has temporarily stabilized.

The actor does not fix the meaning of her actions and, by extension, does not fix the meaning of the life composed of these actions. She is the protagonist of the story of her life but not its sovereign biographer. Arendt writes:

Although everybody started his life by inserting himself into the human world through action and speech, nobody is the author or producer of his own life story. In other words, the stories, the results of action and speech, reveal an agent, but this agent is not an author or producer. Somebody began it and is its subject in the twofold sense of the word, namely, its actor and sufferer, but nobody is its author.[28]

As much as we might wish otherwise, we do not author our lives. Rather, we act in a web of relationships, and like other participants of this network of agents, we respond to actions. That, Arendt insists, is the human condition. And we come to terms with the meaning of all this interaction and the events it precipitates only when, having reflected on our experiences, we put our "story into shape."[29] The meaning of our life arises through the weaving of a story in the wake of our interactive deeds.

The ancient Greeks, at least as far back as the sage Solon, insisted that no man should be called happy until he is dead (and with these words, the tragedy of *Oedipus Tyrannus* concludes). The statement was meant to underline the unpredictability of human affairs. At a deeper level, it suggests that the meaning of an action cannot be ascertained until it has issued in a story. In turn, the identity of an actor and the meaning of her life—who an actor is and what her life signifies—cannot be known until the story that chronicles and makes sense of it gets put into shape. This story, at least in its complete form, cannot be told until death prevents the protagonist from initiating any more acts or responding to the deeds of others.

The meaning of an action, like the meaning of a life, congeals only through a retrospective narrative. But lives do have distinct origins. The story of a life begins at birth. For this reason, Arendt suggests that the phenomenon of natality—the fact that we are born—offers a key insight as to the nature of action.

A birth can always be linked to antecedent events. Every birth, like every action, has its causes. It is rooted in biological processes, social structures, and personal relationships. Despite such causes and contexts, a birth, like every other action, marks a true beginning. It is synonymous with novelty. How so? Arendt's answer is that causes and contexts do not and cannot rob birth of its effect—the initiation of a series of unprecedented interactions. The birth of a human being sets an original tale on its way.

Notwithstanding its preconditions and various causes, then, birth introduces something new into the world. Human action is akin to birth in that it produces beginnings by provoking a unique series of responses. That is why the newness of action (and its meaning) can be ascertained only retrospectively, in a narrative that captures the responses it provokes and its effect on the world. Arendt argues that this opportunity for beginnings inherent in our capacity for action is "guaranteed" by the fact of birth.[30] The human capacity for political action—for beginnings without certain endings—announces itself at every birth and reasserts itself every time we provoke others to respond.

The actor discovers the meaning of her action (and her life) by placing it within a story. But the narrative nature of action is not limited to retrospective accountings. Actors envision and invent alternate scripts as potential avenues for their future. Whether engaged in a retrospective or prospective accounting, actors find themselves inhabiting stories. As John Dewey observed, we employ imagination to forecast ourselves in certain plots and roles, and respond emotionally to this "running commentary of likes and dislikes, attractions and disdains, joys and sorrows."[31] This imagined narrative of the self, which we inhabit with our hearts as well as our heads, instructs our current pursuits. Philosopher Alasdair MacIntyre makes this point succinctly when he writes: "I can only answer the question 'What am I to do?' if I can answer the prior question 'Of what story or stories do I find myself a part?'"[32] The effort to determine what is to be done is not a matter of passively adopting settled narratives. As protagonists, we may manifest traits of inventiveness and ingenuity. Our script may call for spontaneity, creativity, and imagination. In any case, alternate storylines arise for every actor. There is always a choice to be made between competing scripts and roles. Political actors insert themselves into half-told tales—stories that await novel words and deeds and unforeseen responses to take them in new directions.

Political Judgment

How does the political actor make choices between competing scripts and roles? How does the political actor understand and evaluate the initiatives of others and decide on the best response? This is the task of judgment.

Judgment is a political capacity. It is what the virtue of practical wisdom allows. Ever since Plato and Aristotle first discussed *phronesis*, and Cicero, Rome's foremost politician, deemed prudence the greatest virtue, the faculty of judgment has been at the center of theories of politics. Without it, the life of the *zoon politikon* is impossible. The health and welfare of a polity depends on its leaders' exercising good judgment in all affairs of the state: diplomatic, military, financial, legal, and moral. In democracies, citizens actively share in the virtue of practical wisdom. Aristotle deemed participation in judgment and decision making in the pursuit of justice the distinguishing mark of a citizen. To be sure, a free and prosperous nation cannot be achieved or sustained in the absence of a judicious citizenry. While action is the essence of political life, wise judgment allows a good political life. But what does it mean to judge, and to judge well?

Good judgment is an aptitude for assessing, evaluating, and choosing a course of action in the absence of algorithms or principles that determine with certainty the right response. The political world is one of deep complexity: the web of relationships that defines it is intricate and convoluted. This web cannot be well navigated by following rules, notwithstanding the importance of employing them as guideposts. Good judgment, in this respect, is not simply an act of calculation or an exercise in logic. Deductive and inductive reasoning play their part. Still, to judge well is to be responsive to context, including those facets of political life that escape reason. The good judge must attend to the subtle interplay of passion and prejudice, belief and misbelief, complex and contradictory motivations and desires. After all, this is the stuff that informs human action. Employing reason alone, we might be able to predict what wholly rational people would do in particular situations. But we would fail miserably at predicting what actual people will do. So the prudent person, as Cicero observes, must gain access to and assess people's "thoughts and feelings and beliefs and hopes."[33] To judge well is to learn how to read complex and convoluted hearts and minds. Inevitably, one must read between the lines.

Arendt observed of judgment that it was a "political rather than a merely theoretical activity."[34] Judgment is political because it involves us in a network of relations characterized by plurality. Its political nature stems from the requirement that we gain access to the hearts and minds of a diverse population of fellow citizens and statespeople. Arendt calls this "representative thinking"—understanding how one would think, feel, and act were one in another's shoes.[35] Representative thinking demands a rich, nuanced appreciation of how others, given their experiences and points of view, might form judgments and take actions.

Ralph Waldo Emerson wrote, "Prudence is false when detached."[36] Arendt's concept of representative thinking reflects this insight. The ability to understand other people's perspectives and predilections, the ability to walk in their shoes, is a crucial component of good judgment. Representative thinking is an act of empathetic imagination. Without imaginative insight into the deeply complex world of others' hearts and minds, our judgments would be limited to our own beliefs, prejudices, and interests. As such, our judgments would echo personal hopes and fears rather than reflect the actual contours of the political world. As Mark Twain observed, "You can't depend on your judgment when your imagination is out of focus."[37] Failure in judgment is most often a failure of imagination. To judge well is to be able to imagine how others will assess their options, grapple with uncertainty, emotionally react to demands and opportunities, and pursue their interests.

To engage in representative thinking, one imaginatively places oneself in other people's positions. Importantly, these positions are not stationary. Representative thinking requires understanding where people are coming from and where, following this trajectory, they are likely to go. It is the ability to inhabit the dynamic stories of others. It is often said that before you judge someone, you should walk a mile in her shoes. This is wise counsel. To judge anyone well, one must imaginatively accompany that person on her journey.

To develop political judgment requires that one become attuned to the stories of all the relevant actors—the stakeholders in any given context. It requires that one imaginatively inhabit the *oikos* of others. Effectively, to judge is to assess and evaluate a macronarrative in the making, predicting how it is likely to develop and how it should unfold to achieve the best practicable results. Such macronarratives are composed of a plethora of intersecting and embedded micronarratives—the interdependent stories that capture the interactive lives of each of the characters.

The novelist Henry James once observed that "character is plot." James was suggesting that a writer's first task is to create strong characters. The plot of a novel then unfolds as the writer imagines his characters living out their intersecting, personal dramas. Likewise, the statesperson or citizen judges an event or opportunity by assessing how the stakeholders involved, given their respective characters and interests, will interact. She then assesses how the plot of the story will likely develop as a result of these interactions and how she might take initiative, respond, or intervene to achieve the best results.

The political theorist Isaiah Berlin observed that good judgment was an achievement of the "profounder students of human beings."[38] Such students of life are acutely aware of "the infinite variety of the social and political elements in which they live. Their antennae are extremely sensitive and record half-consciously a vast variety of experience."[39] The gifted novelist, Berlin notes, also displays this capacity. Along with geniuses of politics such as Abraham Lincoln and the Prussian statesman Otto von Bismarck, Berlin finds the faculty of judgment best developed in "the great psychological novelists" as well as in historians and dramatists.[40] In each case, the good judge is skilled at investing herself in the real or imagined stories of other people, understanding the world from their perspectives, and acting (or writing) accordingly. François de La Rochefoucauld, the seventeenth-century French author and aphorist, wrote: "To try to be wise all on one's own is sheer folly."[41] Good judgment is a capacity to partake in the company of others. We develop political judgment by entering the hearts and minds of our fellow citizens and statespeople through our imaginative venture into the stories of their intersecting lives.

Action as an End in Itself

It is said that one should judge people not by their words but by their deeds. To be sure, actions are a better guide than opinions. We generally judge actions by their results. The proof, we say, is in the pudding. An action that does not produce anything useful, beautiful, or beneficial is seldom deemed important or worthy. In the world of politics, however, you can never do just one thing. Action, the essential feature of political life, finds its meaning not in the singular intentions or motives of the doer, but in the multiple, and often unpredictable, responses of others. Indeed, an action that did just one thing would not be an action at all. It might mark the creation of a work of art, the crafting of a technological artifact, the writing of a grocery list, or the earning of one's bread by the sweat of one's brow. All of these efforts might reasonably be said to aim at and, for the most part, achieve one thing. Political action is different: it engages a diverse world of unique individuals by soliciting responses. The results of its provocations are always multiple, mixed, and uncertain.

Given this understanding of action, one might expect that politics would necessarily be a tragic, anxiety-prone, and frustrating affair—and often it is. After all, its essential element, action, delivers us into a realm where we, like Oedipus, are prevented from mastering our fates. Consider the statement by Friedrich Engels, who collaborated with Karl Marx to explore the nature of political economy. Engels wrote that historical events are the products of "innumerable forces which interlace, an infinite number of parallelograms of forces giving a resultant, the historical happening. This, in its turn, can be regarded as the outcome of a force acting as a whole, without consciousness or will. For that which each individual wishes separately, is hindered by all the others, and the general upshot is something which no one in particular has willed."[42] Engels and Marx despaired at the prospect of history developing as the unpredictable effects of interactions. They hoped that a communist society would put an end to such history, giving humankind greater control over its fate.

Many of the ideals that Marx and Engels espoused are inspiring. Their egalitarian society is a place where all basic needs are met, human potential is developed, and people are no longer alienated from nature. But as we know from historical efforts to put communist visions into practice, attempts to rid the social realm of unintended consequences by way of

massive state control are catastrophic. They pit a vanguard of elite ideo-
logues in a war against the dynamism and plurality of the citizenry. The
first casualty in this war is freedom. In sacrificing human freedom to the
effort of ridding the world of unintended consequences, communism
throws the baby out with the bathwater.

Aristotle and Hannah Arendt took a different course. They understood
the social and political world to be a place rife with unintended consequences. A world where you can never do just one thing is always a potentially tragic place. But when all you have are lemons, the best thing to do
is to make lemonade. Aristotle and Arendt effectively make lemonade: they
deem action to be its own reward.

The political actor can neither achieve certain results nor control the
meaning of his deeds. But that is no reason to despair, according to Aristotle and Arendt, for action is essentially self-fulfilling. Unlike laboring to
earn one's keep, crafting a piece of technology, or painting a picture, action
does not find its redemption in a final product (such as a wage, a technological artifact, or a piece of art). Action is not a mere means to achieve a
particular end. Rather, it is an end in itself.

Aristotle was the first person to articulate and justify this astounding
claim. He maintained that there was a distinct difference between a craft-
sperson who makes something and a citizen who does something. Craft
production or fabrication was what the Greeks called *poiesis*. Citizens and
statespeople, in contrast, were engaged in something called *praxis*. *Poiesis*
is aimed at fabrication. Its effort is justified by way of that which it produces. In contrast, *praxis* does not aim at some identifiable end, and hence
it cannot justify itself by faithfully realizing a plan or duplicating concretely a model held in the mind's eye. Aristotle stipulates that "production
has its end beyond it; but action does not, since its end is doing well
itself."[43] In this respect, *praxis*, or political action, has the characteristic of
a virtue. Just as virtue is said to be its own reward, so action finds its justification and redemption not in what it finally achieves but in its being
done well.

Action cannot find its justification or redemption in a final achievement
because no one really knows when its work is done. An action enters the
political world as a solicitation. The responses it provokes, and the actions
these responses in turn stimulate, may continue to ripple across the web
of a political community indefinitely. As noted earlier, when Martin Luther

King Jr. spoke of his dream for a righteous America beneath the statue of Abraham Lincoln, he provoked an outpouring of responses. His words, and the ripples they sent across the United States and the rest of the world, continue to draw forth responses today. While many stories have been told to capture the meaning of King's action, its political repercussions have not wholly abated. There will always be more stories to tell. For Aristotle and Arendt, action must be engaged for its own sake rather than for the effects it produces because the actor does not and cannot control the effects of his action. Political action is not in the business of delivering final products. It is in the business of beginning something well.

Plutarch, the ancient Greek philosopher and biographer, observed, "They are wrong who think that politics is like an ocean voyage or a military campaign, something to be done with some end in view, something which levels off as soon as that end is reached. It is not a public chore to be got over with; it is a way of life."[44] Politics, for Plutarch and Aristotle, was a way of life whose chief component, action, was its own reward. The ancient Greeks understood politics as the achievement of citizens who managed to escape tyrannical rule so that they might organize themselves freely in pursuit of the common good. This pursuit well carried out—not a final destination arrived at or goal achieved—allowed the practice of virtue and the good life.

Ironically, the terms *politics* and *political* are often used today to suggest quite the opposite. Rather than referring to the virtuous and self-fulfilling pursuit of the public good and the good life, *politics* generally connotes the strategic and manipulative effort to serve the narrow interests of a particular individual, group, or party. Abraham Lincoln reflected a popular sentiment when he characterized politicians as "a set of men who have interests aside from the interests of the people, and who, to say the most of them, are, taken as a mass, at least one long step removed from honest men."[45] With similar sentiments, the American satirist Ambrose Bierce defined politics as "a strife of interests masquerading as a contest of principles. The conduct of public affairs for private advantage."[46] This cynical understanding of politics is neither particularly modern nor particularly American. The ancient Greeks already spoke of politics in such pejorative terms.

It is precisely because politics so frequently deteriorates into the conduct of public affairs for private advantage, and is populated by partisan politicians "at least one long step removed from honest men," that theorists from Aristotle to Arendt have found it necessary to reassert its potential. Once political action ceases to be understood as its own reward in the virtuous pursuit of the public good, it becomes at best a purely administrative or technical activity. And once we demand that action, like technical fabrication, "obtain a pre-determined end," Arendt observes, it will "be permitted to seize on all means likely to further this end."[47] At that point, politics becomes a war by other means, winner take all. When in the organization of our collective life the ends are understood to justify any and all means, we are one step away from tyranny.

Of course, we could define political life as an instrumental endeavor much like craftsmanship or economic pursuits without giving up on democratic safeguards against tyranny. Indeed, many political scientists base their theories of democracy on economic models that assume action is purely instrumental. The polity is portrayed as an electoral marketplace. Citizens spend their votes—well or foolishly—in an effort to satisfy their own particular interests. Effectively they trade their ballots in anticipation of public policies that will meet their needs, policies that politicians pledge to enact if elected. In turn, candidates for office effectively sell themselves to citizens during their campaigns. Like any salesman, they want to get the most they can from their customers while offering the least in return. So they promise only as much as they must to get the votes they need, and then they deliver even less once in office. Of course, their legislative efforts will have to be sold to the electorate during the campaign for reelection. So as soon as they gain power, politicians begin the task of repackaging themselves in an extensive advertising campaign. The market cycle begins anew.[48]

In a political marketplace, the actions of citizens and statespeople are simply means to attain particular ends. Their decisions about when and how to act are made on the basis of a self-interested, cost-benefit analysis. The value of the end (the benefit received) must justify the means (the aggregated costs of one's efforts) before action is taken. For all its analytical appeal, such economic models of democracy fail to capture a peculiar feature of political action.

The problem with the instrumental model is that many of the actions that should be considered the costs in political life are actually perceived as benefits by the actors themselves. For example, the economic model of democracy posits the time and effort of going to the polling station and waiting to vote as a cost. It is the price people pay to increase their chances of electing a candidate who will best serve their interests. But citizens who vote often proudly describe this civic duty as rewarding and fulfilling. Voting is experienced as a benefit, not a cost. The same might be said about many legislative efforts, which far from being experienced merely as costs to be borne by politicians seeking reelection, are perceived as rewarding and fulfilling acts of statesmanship.

Economist Albert O. Hirschman has suggested that political life is much like a pilgrimage in this regard. Religious pilgrims often journey long distances and suffer significant hardships to reach the shrine where they will worship. The arduous journey might well be considered a cost to be borne such that the benefit of worshiping at the holy site can be gained. But as Hirschman points out, the trials and tribulations confronted during the voyage are intrinsic parts of the experience that pilgrims seek. Indeed, long distances to be traveled and hardships likely to be borne often serve as a stimulant rather than a deterrent.[49] Hirschmann insists, as do Aristotle and Arendt, that political action is not reducible to an instrumental effort to achieve an external good. It is, at least in part, an end in itself.

We often speak of the "courage to act." Action demands courage in a way that administration and technical fabrication do not. Courage is demanded of the political actor because she does not control the responses to her actions or its meaning. In this respect, political action is always heroic. Arendt writes that "the word 'hero' originally, that is, in Homer, was no more than a name given each free man . . . about whom a story could be told. The connotation of courage, which we now feel to be an indispensable quality of the hero, is in fact already present in a willingness to act and speak at all, to insert one's self into the world and begin a story of one's own."[50] While an action begins a new story, it does not determine the plot, the cast of characters, the conflicts and resolutions, the climax or denouement. The political actor courageously involves herself in a dramatic performance that she does not control.

In the Bhagavad Gita, the great Hindu epic poem, we read of divine Krishna's instruction of Arjuna, a prince who finds himself confused and immobile in the face of moral tensions and political roles. The god urges Arjuna to act but also offers this counsel:

You have a right to your actions, but never to your actions' fruits. Act for the action's sake. . . . Without concern for results, perform the necessary action; surrendering all attachments, accomplish life's highest good. . . A man deluded by the I-sense imagines, "I am the doer. . . ." The sage, wholehearted in the yoga of action, soon attains freedom . . . he is unstained by anything he does. The man who has seen the truth thinks "I am not the doer."[51]

Krishna counsels Arjuna to display the courage to act, to become involved in a dramatic performance that he does not and cannot control. Arjuna should not forgo action, only mastery of the results of action.

Whatever intentions and goals actors bring to the stage of politics—however instrumental, strategic, or self-serving their efforts may be—the significance of what they do is not theirs to determine. An action prompted by the most strategic and self-serving of motives, as many if not most political actions are, will still produce unintended consequences. It will become an element of a story whose meaning the strategic, self-serving actor cannot determine. As political actors, we are not directors of our own drama. Rather, we reveal ourselves as participants in a collective performance. The actor, while often the heroic agent of a story, is not its author. Notwithstanding this rebuff to the human pursuit of mastery (or perhaps because of it), Arendt, like Aristotle, maintains that the greatest achievement that human beings are capable of is to be found in the living deed that manifests human freedom and exhibits virtue in a life well lived.[52]

Aristotle held that democracy entails ruling and being ruled in turn. The democratic exercise of power implies not only acting and effecting change, but also being acted on and undergoing change.[53] That is the effect of the interdependence citizens experience in the political *oikos*. Political action engages one in a process whereby goals and interests, and consequently values and identities, become shaped and reshaped, formed and transformed. Like a participant in a dialogue, the political actor involves himself with others in an interactive exchange that will yield unforeseen conclusions, new relationships, and, most important, new self-understandings and identities.

Political theorist Joseph Dunne, describing Aristotle's theory of politics, writes that the actor

is constituted through the actions which disclose him both to others and to himself as the person that he is. He can never possess an idea of himself in the way that the craftsman possesses the form of his product; rather than his having any definite "'what" as blueprint for his actions or his life, he becomes and discovers "who" he is through these actions. And the medium for this becoming through action is not one over which he is ever sovereign master; it is, rather, a network of other people who are also agents and with whom he is bound up in relationships of interdependency.[54]

In a similar fashion, Hannah Arendt holds that action discloses not "what" we are but "who" we are. This "who" is not master of his world, not a sovereign entity, but part of a "web of human relationships."[55] Acting and responding to others' actions within this web of relationships allows the experience of freedom and the self-fulfilling practice of virtue.

Conclusion

Politics concerns the pursuit of public goods. Sustainability is such a good. Its pursuit requires that power, the central political resource, be shared. Without an equitable distribution of power, the equitable distribution of other goods could not be achieved or sustained. Political action is crucial to the pursuit of sustainability. But politics is more than a mere means to a desired end. It is, like sustainability itself, not a destination to be reached. Rather, it is a process to be engaged well, a worthy voyage to be virtuously undertaken.

The political world is defined by its plurality, by the diversity of unique individuals who compose it. In such a world, no action can have a singular, predictable effect. That is true whether the action in question is as commendable as the quest for social justice or as self-serving as the pursuit of individual gain. Whatever intentions actors bring to the stage of politics, however strategic and self-serving their efforts, the significance of what they do is not theirs to determine. Political life is a story told by many, filled with initiative and indeterminacies, open to interpretation, and manifesting freedom.

It would be reassuring to believe that the impact and meaning of actions were ours to control. We all want to be masters of our fates. But that is not our lot. That is not the human condition. And the reason is simple: we can never do just one thing. Whenever we act, we are acting in a web of relationships. Our actions send ripples through this web, producing multiple effects and side effects. We can be virtuously responsive to this rippling web of relations, but in the end, we cannot control it. In such a world, it is better to enjoy lemonade than be left with a sour disposition.

Action engaged for its own sake is but a part of politics, though an essential part. The pursuit of power, the thrust of ideology, partisan machinations, electoral strategies, logrolling, bargaining, compromise, rhetoric, pandering, and manipulation—this is the mundane stuff of politics. To highlight the capacity of action to be its own reward is not to suggest that political actors do not engage in these strategic and purposeful activities or are not motivated by external goods and self-interest. Political actors are not all whimsy and spontaneity; instrumental, self-serving effort is certainly not foreign to their nature. Political actors without purposes do not exist: self-serving motivations are intrinsic to virtually every political endeavor. But political action, to merit its name, must display a self-fulfilling, interactive, noninstrumental quality. Otherwise the behavior in question might well be a form of administration, or management, or tyranny, or technical proficiency. But it is not politics.

Politics occurs whenever we muster cooperation and mitigate conflict to a degree that makes the identification and pursuit of common goals possible. In turn, politics occurs whenever we come to define or redefine ourselves as members of a community capable of such common pursuits. Political action allows new and unforeseen events to occur, and new and unforeseen perceptions and identities to emerge. In a democracy, this fact not only describes the nature of political action; it is meant to inform the self-understanding of citizens and statespeople.

In political life, the effects of our actions always escape our control. This limitation is often viewed as an affliction. To be sure, it can be frustrating, even debilitating. At the same time, the fact that we can never do just one thing when we act politically allows us to experience freedom. The freedom inherent in political action does not depend on the actor's ability to sever a causal chain of events. Neither does it depend on the originality of an actor's motives or intentions or even on how effectively these desires or acts of will are fulfilled. Rather, freedom is experienced in the provocation of an unpredictable set of responses. In the realm of politics, freedom is not found in sovereignty. A sovereign independence from community might be the best way to describe the freedom of omnipotent gods. Human freedom can be discovered and manifested only by way of interdependence.

As political animals, we face the challenge of cultivating and shaping our interdependencies so as to increase the benefits of collective life and mitigate the inherent problems of collective action. To no small extent, this requires reducing the insecurities and uncertainties that plague the human condition and constrain the fuller development and cooperation of its members. In turn, the challenge is simultaneously to cultivate the freedom for uncertainty that political action demands and manifests. This is not to abandon responsibility for one's deeds. Far from it. Awareness of the web of relationships that we act within but cannot control might well heighten our sense of responsibility. Minimally it should prompt us to refrain from seeking mastery in all things. Such a chastening is in order lest the human race as a whole suffer the fate of Oedipus.

6 Psychology

The psyche is the greatest of all cosmic wonders.
—C. G. Jung

This chapter explores the nature of the mind. Its basic thesis is that an individual's participation in and sense of belonging to the web of life is best grounded in her appreciation of the interdependent parts of the mind—that inner *oikos* sometimes called the soul, psyche, or, simply, the self. In chapter 5, we saw that freedom is experienced in the realm of politics through the responses that action provokes within a web of interdependence. Likewise, we experience freedom of the mind or soul through an inner, psychological interactivity.

Though an internal dwelling place, the mind has a powerful impact on the external world. Look around, and you will see a planet transformed by the human mind. Feats of science and engineering, industry and commerce, politics and the arts: all of these products of the human mind have reshaped its worldly habitat. The mind not only transforms its earthly *oikos*; it also transforms itself. By redirecting its powers, the mind can alter its own strength, flexibility, speed, endurance, and focus. Indeed, the human mind can transform the very brain and body that house it.

The power of mind over bodily matter is clearly demonstrated by the placebo effect. Before new pharmaceuticals are put on the market, they are tested against medically inert substances called placebos. In studies designed to gauge the effectiveness of the new pharmaceuticals, some participants unknowingly take a sugar pill rather than the drug. Subsequently the improvements in the health of those taking the drug and the placebo are measured against each other. This is done because the human mind has the power to promote physical healing. If people believe that they are getting a drug that will cure their ailments, they often regain health, or demonstrate reduced symptoms, even when the treatment they receive is completely fake. The percentage of patients demonstrating improvement from placebos ranges as high as 40 to 50 percent, though it is usually somewhat lower.

Placebos empirically demonstrate the so-called power of suggestion. The belief that a pill will decrease allergy symptoms, or that surgery will alleviate symptoms of Parkinson's disease, actually reduces allergy symptoms and alleviates symptoms of Parkinson's disease notwithstanding the fact that the pill contained no active ingredients and the surgery was an orchestrated sham. In such cases, what people believe to be true makes it true. Often placebos can be effective without patients' registering any false beliefs in their curative powers. They can be the product of unconscious associations or "conditioning."[1] The mind is often posed as distinct from the body. The placebo effect demonstrates that mind and body are interdependent.

The human mind is a powerful thing. It allows us to transform ourselves and our world. At times, owing to its very power, the mind gives the false impression that it is autonomous and independent. The Iroquois peoples were aware of this tendency. To combat it, they spoke of themselves as having a *longbody*: a network of connections between the self and other members of one's tribe, both living and dead. A longbody is also a network of connections between the self and its world. For the Iroquois (as well as the Hopi, Navajo, and Pueblo peoples of the American Southwest), the experience of self or soul is not limited to the mental thoughts, feelings, and images that we usually call the mind. Rather, they experience the self or soul as an extended web that includes living and dead family and tribal members, animals eaten, things possessed, and tribal lands. A longbody consists of all the strands and connections within the web of life that make possible the specific node called the *self*.

In a genetic sense, there is no denying that we have longbodies. Our gene pool connects us quite literally to other family and tribal members, living and dead. Genetically we are composed of this ancestral legacy. Indeed, our gene pool connects us to virtually every other form of life that evolved. We share DNA with the simplest of cells, and our physical bodies are intimately linked to the web of life in other ways. Diverse forms of life do not simply live alongside us: they exist inside us. Indeed, there are more cells that belong to bacteria and other organisms in our bodies at any given moment than cells carrying our own DNA. We host a web of life within us. As evolutionary biologist Richard Dawkins puts it, "We are all symbiotic colonies of genes."[2] Many of the genetically diverse organisms that inhabit us, such as the bacteria living in our stomachs and intestinal tracts that aid digestion, are crucial contributors to life functions.

Our physical bodies are connected to the web of life in yet another way. Over the span of a lifetime, the average human being consumes somewhere between 60,000 and 100,000 pounds of food. All that food goes into the making, repairing, and replacing of the cells that compose our bodies. We are, quite literally, what we eat. Beyond food ingested, we inhale about a quadrillion molecules of air with every breath, air that has been circulating within the web of life for eons. Students taking biochemistry exams are often asked to confirm mathematically the fact that each breath they take, on average, contains tens of millions of molecules that were once exhaled by Mozart—and by Attila the Hun. In his "Song of Myself," Walt Whitman wrote that "every atom belonging to me as good belongs to you."[3] It was both good poetry and good biochemistry. Like everything else around us, we are composed of atoms that were created billions of years ago when the stars were born and have been combining, dispersing, and recombining ever since. We are but their transitory hosts.

At genetic, molecular, and atomic levels, our longbodies also extend us into the future. Our genes get passed on to our children, and from them, to a diminished degree, to grandchildren and great-grandchildren. The food passed through our bodies becomes nourishment for other plants and animals, just as our physical bodies themselves will provide nutrients to various organisms when we die and molder. Every atom in our bodies will be recycled into other organic and inorganic forms once we shuffle off this mortal coil. The embodied self is merely a temporary placeholder for a longbody that extends from the ancient past into the distant future. The Iroquois did not know about the genetic, molecular, and atomic means by which selves are connected to the web of life. But in speaking of a long-body, they knew what they were talking about: interdependence. *Mitakuye oyasin*—we are all related.

Reading Minds and Staying Connected

It is said that insanity is best identified by the afflicted person's sense of separation from the rest of humankind. To be mad or insane, the height of psychological disorder, is to be unable to feel connected to other human beings. A complete loss of contact with social reality is known as the state of psychosis. To be psychotic is to be in a world of one's own.

Connecting with other minds does not entail losing sight of oneself as a unique individual. We experience ourselves as distinct selves. That is crucial for mental health. Were the experience of individuality to be wholly absent or severely diminished, other psychological disorders would be in play. One might be facing the disorder of megalomania, which subsumes the world into the ego, or radical codependence, where the autonomous self evaporates. Mental health constitutes a sense of connection to other minds that is not, at the same time, a wholesale merging or amalgamation. Our sanity, and beyond that our psychological well-being, is maintained through the linkage between the individual *psyche*, the Greek word for mind or soul, and a world of other *psyches*. In a word, psychological well-being is grounded in the experience of interdependence.

Verbal communication allows us to understand what is going on inside other people's heads. When a person "speaks his mind," he allows us access to his thoughts, beliefs, and feelings. We come to see him as a distinct individual and, at the same time, someone with whom we share a great deal. Language is a fundamental means of realizing psychological interdependence. But it is not the most fundamental. Our ability to share ourselves and our world through the spoken or written word is grounded in an even more basic means for gaining access to each other's hearts and minds. Human beings are mind readers.

To say that we can read others' minds is not to suggest that some sort of parapsychology or extrasensory perception is at work. Rather, *mindreading* is the term evolutionary psychologists have adopted for the human capacity to recognize other minds and discern what the world is like from their points of view. The power and sophistication of our ability to read minds distinguishes us from the smartest of our evolutionary cousins, the apes. Indeed, mindreading might be considered the capacity that most distinguishes humans from other animals. Very early in life, we develop an understanding that people act in reference to their own states of mind.

That is, we interpret people's actions as products of their beliefs, intentions, and goals.

Of course, many other animals, and particularly other mammals, are quite adept at predicting behavior (and adjusting their own) based on observation. Think of predators as they stalk or chase prey. But when other animals respond to observed behavior, they do not rely on an interpretation of mental states. They simply respond to behavioral patterns. Only the great apes come close to the mindreading ability of humans, and at best they achieve the level demonstrated by a six-year-old child. Older children go on to develop a much more sophisticated capacity to predict the behavior of other individuals, including the reactions of individuals who hold false beliefs. This is something no other animal can do.[4]

Consider the following experiment. A couple of four-year-old children, Jack and Jill, are shown a cookie that is then hidden in a cupboard. While Jill turns her back, the cookie is moved to a breadbox. Jack is then asked where Jill will look for the cookie. He responds, incorrectly, that she will look in the breadbox. The boy ignores the false belief inside Jill's head. In this respect, young children are effectively "mind blind." In a couple of years, that will change. An older Jack will realize that Jill will look for the cookie in its original spot because she is unaware that the treat was moved while her back was turned. Older children effectively learn to peer inside the minds of people they interact with. Their ability to sift through others' correct and false beliefs allows them to predict behavior better.

How do we gain access to the minds of others? In the early 1990s, neuroscientists employing brain-imaging technology discovered that what was going on inside the head of a monkey as it picked up an apple was virtually identical to what was going on inside the head of a monkey that was watching another monkey pick up an apple. Studies with humans soon corroborated these findings. The brain cells that are activated in such circumstances are called *mirror neurons*. Mirror neurons become active whenever we engage in an action. To a lesser but still significant degree, these same neurons become active whenever we see another person engage in an action.

Evolutionary biologists speculate that mirror neurons developed among social animals that have relatively long developmental periods during which the young are cared for by parents or other community members. Among such animals, mirror neurons facilitate learning by example.

Because doing an action and watching it being done get the same neural networks fired up, the cerebral work required to master a particular skill could happen twice as fast. Monkey see, monkey do, or, rather, monkey see often, monkey do better.

In human communities, the same is true, only more so. Human beings have many more mirror neurons than other primates and mammals. The rapid development of human culture is hypothesized to be largely a product of them. Humans can learn a wide range of skills and develop such an expansive behavioral repertoire owing to their innate capacity for imitation. Even the development of verbal communication appears to be the product of mirror neurons. Here, so-called echo neurons facilitate sound and speech imitation in the same manner that visually oriented mirror neurons facilitate the imitation of motor activity.[5]

Mirror neurons foster learning by example. Importantly, the examples need not actually be observed. Calling them up in the mind's eye can do the trick. Simply imagining certain actions being performed can significantly increase a person's ability to perform these actions because the skills required rely on the same neural pathways. Participants in one study imagined shooting basketball free throws each day for a specified time. They improved their skill levels almost as much as did participants who actually practiced shooting baskets for the same amount of time. Other studies have demonstrated that both mental and physical aptitudes can be developed and sharpened merely by imagining their exercise.[6]

Despite their name, mirror neurons do not only help us mimic action. They also help us understand the intentions or goals behind actions. A person watching the idiosyncratic movement of another person's hands will, thanks to mirror neurons, be able to imitate that movement. When the movement has a clear intention behind it, however, mirror neurons in the imitator will not fire in mimicry of the idiosyncratic movements. Rather, they will fire in a fashion that mimics the action needed to achieve the intention. When we watch someone awkwardly pick up an apple and bite it, for example, our mirror neurons fire in the same way as were we picking up the apple and biting it. The awkward movements get ignored. In other words, our mirror neurons do not simply mimic the movements of those we observe. They allow us to understand the purpose of these movements and help us achieve it.

Neuroscientists maintain that as you watch someone engage in a particular activity, your brain, with mirror neurons firing, acts as if you are performing these same actions. In turn, another part of your brain interprets the activity of the active mirror neurons, assessing the intention behind the action. Then your brain projects this assessment onto the active person being observed. This neural process allows you to pursue similar goals or, alternatively, to predict the future behavior of others based on assumed motivations. Your brain adopts a "simulation routine" and interprets it to gain access to the world of others.[7] Effectively you come to understand the mental states of other people by simulating their neural activity inside your own head, and then interpreting this simulation. Of course, all of this simulation and interpretation is going on without conscious effort. As neuroscientist Marco Iacoboni writes, "mirror neural activity . . . reflects an experience-based, pre-reflective, and automatic form of understanding other minds."[8] One might say that humans are hardwired for mindreading.

Reading minds is not melding minds. Mirror neurons demonstrate our psychological connection to, not our mental union with, other people. The activity of mirror neurons is more pronounced when we perform an action ourselves than when we watch that action being performed. Our brain clearly distinguishes between our own actions and those of other people. We exhibit psychological independence. At the same time, our mirror neurons fire when we observe other people acting or speaking, whether we like it or not. This underlines our dependence on others. The mind's simultaneous independence from and dependence on other people bespeaks our psychological interdependence.

To understand other people well, we must gain access not only to their minds but also to their hearts. As I noted in chapter 5, the representative thinking and practical wisdom that allow us to be skillful social and political animals does not restrict itself to gaining awareness of other people's cognitive states of mind. Good mindreaders understand not only what people think, believe, and intend but also what they value and feel.

Mirror neurons are at work whenever we connect to others emotionally.[9] Just as we have evolved to understand and interpret intentions and beliefs, so we have evolved to understand and interpret feelings. As with intentions, the understanding and interpretation of emotions occur by way of an inner simulation. When you observe someone in pain or glee, the corresponding mirror neurons start firing in your own brain. This inner simulation, though not as pronounced a neuronal firing as would occur if you experienced suffering or joy directly, allows you to understand and interpret the feelings of other people. An emotional connection via mirror neurons occurs without verbal communication or conscious effort. As Marco Iacoboni writes, "Mirror neurons . . . demonstrate that we are wired for empathy. . . . We are deeply interconnected at a basic, pre-reflective level."[10] In short, we cannot help but feel for others.

Smile, and the world smiles back at you. Laughter, we know, is contagious. When smiles and laughter are shared, more than physical movements are being imitated. The emotion behind the physiology also proves infectious. Joy or gaiety is actually experienced by those witnessing, and imitating, smiles or laughter. One can even infect oneself in such manner. The act of willfully smiling can, at least temporarily, lift one's mood. Likewise, the act of clenching one's fists can stimulate anxiety or anger. Such stimulated emotions can affect beliefs and judgments. In one study, people were asked to evaluate applicants while engaging in certain movements. Some of the subjects were told to carry out their evaluations while moving their arms in a manner typically associated with pushing things away. Their evaluations proved to be quite critical. Other subjects carried out their evaluations while moving their arms in a manner typically associated with pulling things toward themselves. They provided more positive assessments of the applicants.[11] Just as we can improve a skill simply by imagining its performance, so we can experience a feeling or acquire a mood simply by performing an action associated with that feeling or mood. For good or ill, beliefs and judgments typically associated with these feelings or moods will also be fostered.

Mirror neurons appear to be responsible for this inner drama. Sensing its mirror neurons for the clenching of fists firing, the brain interprets this activity as evidence for anger. Perceiving an assumed effect, the brain invents a cause. It then interprets and reacts to its world based on the assumed reality of this cause. One might imagine the brain saying: "Look at my clenched fists! I must be very angry. Yes, now I feel it. I am angry! Well, I must be angry for a reason. Something out there surely deserves my wrath. Let's see if I can find it." The brain effectively creates an inner feedback loop, producing and reinforcing emotions, beliefs, judgments, and potentially actions.

There appears to be an evolutionary explanation for this tendency of human beings literally to interpret themselves into emotional and cognitive states. In organisms without central nervous systems such as amoebas, the physical body reacts directly to external stimuli. The pseudopod of an amoeba, for instance, will retract on its own when it encounters a saline environment. Because there is no central nervous system or brain to coordinate reactions, local stimulation of the organism produces a local reaction in the organism. Organisms with very basic nervous systems effectively monitored what their body parts were doing in (local) response to environmental stimuli and used this information to figure out what the external world must be like. With the further development of nervous systems, reactions increasingly became less locally controlled. The brain took over. But it still retained the habit of monitoring and interpreting what its body parts were doing, even though its own neural circuits were largely, if not wholly, responsible for what the body parts did in the first place.[12]

Smile for no reason, and you will experience a brief but noticeable elevation in mood. Your brain, perceiving its mirror neurons for a local bodily reaction (a smile) firing, assumes that the world out there must be such as to merit your happiness, and it accordingly stimulates this mood. But because the world does not reinforce this message, your mood quickly returns to the status quo ante. See someone else smile, and the same mirror neurons fire inside you as if you yourself were smiling (though to a lesser degree). The brain goes through the same interpretive activity, with similar results. That is why seeing people smile or hearing them laugh can make you temporarily joyful. Of course, pain and suffering can be as contagious as smiling and laughter are, as the tears and groans of patrons in cinemas well attest. The brain is always interpreting the body and its own neural activity. It provides, monitors, and responds to its own feedback loop.[13]

The brain's capacity to interpret the body and itself helps us navigate our world more effectively. It also provides a sense of self, the core of our psychological reality. Everything from our instincts and habits to our memories and beliefs finds its physiological home in neural networks. The sense of being a "self"—a self that feels relatively stable and unified—arises as the product of the interaction of these networks. Neuroscientist Joseph LeDoux writes that "your 'self,' the essence of who you are, reflects patterns of interconnectivity between neurons in your brain."[14] The self might be thought of as a multilayered neural map. It is the master neural map, composed of all the other neural networks that we inherit at birth or develop over our lifetimes. These complex, intersecting neural networks provide a sense of self that is relatively stable, though amenable to revision based on new stimuli and our own capacity for self-monitoring, reflection, and growth. While the neural networks involved in creating a sense of self are multiple and dynamic, they are, in effect, "locked up in a room together."[15] The "room" in question is the embodied mind, the self that we return to in our reflective moments, hear when we speak, and see in the mirror.

Most of the brain's (self) interpretive work occurs in its left hemisphere. As neuroscientist Michael Gazzaniga observes, the left hemisphere integrates our sensations, perceptions, and bodily reactions to these sensations and perceptions in order to create a "running narrative of our actions, emotions, thoughts, and dreams."[16] This running narrative, which largely occurs under the radar of conscious thought, provides the glue that holds our self, our psyche, together. When we experience or describe our selves, we are experiencing or describing the protagonists of these running narratives. Effectively the psychological self is the product of a mind reading itself as it navigates its world. The self is the protagonist of the mind-built narrative that captures this drama.

All this is to say that our capacity to stay psychologically connected to other people has the same neural foundations as our capacity to stay connected to ourselves. It is the product of an interconnected set of neural maps, a mind that naturally and largely unconsciously imitates, simulates, and interprets its inner and outer world. Psychological well-being arises from the sense of connection we have to the other minds that inhabit our world and to the protagonist of the running narrative that is our life.

Ruling a Divided Soul

In his novel *Steppenwolf,* Hermann Hesse portrays the tormented experience of Harry Haller, a man torn between his wolf-like instincts and his rational, moral conscience. Harry is mistaken, Hesse observes, to believe himself so bifurcated.

A man, therefore, who gets so far as making the supposed unity of the self two-fold is already almost a genius, in any case a most exceptional and interesting person. In reality, however, every ego, so far from being a unity, is in the highest degree a manifold world, a constellated heaven, a chaos of forms, of states and stages, of inheritances and potentialities. It appears to be a necessity as imperative as eating and breathing for everyone to be forced to regard this chaos as a unity and to speak of his ego as though it were a one-fold and clearly detached and fixed phenomenon. Even the best of us shares the delusion.[17]

Harry is complemented for realizing that he does not have a unified, homogenous self. But to think of himself as a soul divided between reason and instinct is too simplistic.

The self is composite. Hesse observes that the separation of the supposedly unified ego into multiple parts is often what passes for madness.[18] But the problem is not that the soul has many parts. The problem with the madman is that these parts do not well interact. Psychological health is grounded in the capacity both to acknowledge and integrate the diverse parts of a multiple soul.

The notion that the mind, self, or soul is not a single entity but an amalgamation of interdependent parts is as old as human self-consciousness. Indeed, the very concept of self-consciousness presupposes a mind that is minimally two-fold. It is a mind that can look back or reflect on itself. Self-consciousness likely evolved as the product of a bicameral (two-sided) brain. The left and right cerebral hemispheres, having taken on separate tasks for efficiency, gained the capacity to monitor each other, with the left hemisphere doing most of the interpretive work. The bicameral brain is clearly one of the major thoroughfares of self-consciousness. But the brain is replete with countless neural networks, many of which cross the hemispheres. Neuroscientists are only beginning to understand how particular networks interact with, monitor, and regulate each other.

The ancient Egyptians understood the soul to be divided into five parts: the *Ren*, *Ba*, *Ka*, *Sheut*, and *Ib*. These five parts correspond, respectively, to a personality, an emotional center or heart, a shadow, a life force that persisted only as long as the person was alive, and an immortal core. The ancient Greeks conceived the soul as divided in three parts, with reason ruling over both spirit and desire. In the *Phaedrus*, Plato famously described the human psyche as a chariot drawn by powerful winged horses. The charioteer holding the reins is the conscious, reasoning intellect (*logos*). The two horses represent the spirit or passion (*thymos*) and the desires or appetites (*eros*). Left to their own, passion and desire will pull a person helter-skelter, producing results destructive to the world and the self. But when passion and desire are well directed by reason, Plato argued, they prove crucial allies. When reined in or spurred on as opportunity and prudence dictate, they can take the soul to great heights.

In a divided soul, some parts are available to scrutiny and direction while other parts operate without conscious awareness or control. The Austrian psychologist Sigmund Freud (1856–1939) is often credited with the discovery of the "unconscious." Freud, the creator of psychoanalysis, famously divided the soul into three parts: the id, which represents instinctual drives; the superego, which represents (moral) conscience understood as the internalized norms and moral strictures of society; and the ego, which attempts to integrate the instinctual drives of the id with the normative dictates of the superego. Conflict naturally ensues in this three-part soul, with the id seeking instant gratification of its desires, the superego laying down constraints and proposing unachievable ideals, and the ego caught in the middle. The inevitable result of this internal conflict, Freud observed, is neurosis.

The problem is that only the ego, with its commonsense effort to satisfy desires within social and moral strictures, is largely available to consciousness. The instinctual drives of the id are shrouded in the unconscious, revealing themselves only occasionally in slips of the tongue and, most powerfully, in dreams, the "royal road to the unconscious." Likewise, the superego remains largely clandestine. Constituted by the internalized norms of father figures and, more generally, of society at large, the superego typically surfaces in feelings of guilt when norms are violated.

Although Plato and Freud both propose the soul to be divided into three parts,[19] their psychologies effectively remain dualistic. On one side are reason, intellect, and conscience. In stark opposition stand the instincts, passions, and desires. Much of the Western tradition of thought, which took its cue from Plato, argues for the victory of reason over passion. Freud suggested that a complete victory was impossible. There would always be inner conflict and neurosis (and hence lucrative jobs for psychoanalysts). The unconscious could never be wholly controlled.

Scooping Freud by more than a decade, the German philosopher Friedrich Nietzsche insisted that most of what goes on in the human mind, including the interminable struggles among its various parts, remains unavailable to consciousness. But Nietzsche, unlike Freud and in direct criticism of Plato, argued that the battle lines in the soul should not be drawn between reason, on one side, and passion, desire, or instinct, on the other. The human soul, Nietzsche asserted, is a multiplicity.[20] It is a conglomerate of passions, desires, and drives. The skirmishes between the multiple parts within this composite self constitute the human drama.

Although trained as a philologist and celebrated as a philosopher, Nietzsche often called himself a psychologist. He dubbed himself a "nut-cracker" of souls and subtitled one his later works *Out of the Files of a Psychologist*.[21] Nietzsche wrote, "For the longest time, conscious thought was considered thought itself. Only now does the truth dawn on us that by far the greatest part of our spirit's activity remains unconscious and unfelt."[22] What surfaces to consciousness—our thoughts, beliefs, and judg-ments—are simply the observed effects of an inner struggle occurring below the register of awareness, a struggle of various emotions and drives, each of which has its own "lust to rule."[23]

Human beings are animals whose instincts have multiplied and broken free of a static, species-wide structure. In the absence of a permanent and stable order, the human soul requires an internal politics to organize itself. Cooperation among the parts is required, as is strong leadership. Soulcraft, for Nietzsche, was indeed a form of statecraft.[24]

The need to create order out of a potential psychic free-for-all is both the distinguishing mark of humanity and the primary cause of its woes. It is the human disease. Nietzsche labels our species the "sick animal," as human beings are more "uncertain, changeable, indeterminate than any other animal."[25] But the disease of our divided state is also the opportunity for human greatness. Nietzsche urged us to embrace the politics of the soul, with all its messiness. Strong leadership is required, but the point is not to create a tyranny. Rather, the soul should reflect the give-and-take of political competition and rivalries. In this regard, Nietzsche might well have seconded Aristotle's recommendation that the type of rule needed in the divided soul was constitutional rather than despotic.[26] The point is to integrate rather than suppress or extirpate passions, desires, and drives. Effectively, Nietzsche recommended for the composite soul what the founding fathers recommended for the thirteen American colonies: to found and maintain a confederation. The health and strength of this "perpetual union" could be achieved only by recognizing the relative autonomy and contribution of each of its constituent parts.

Nietzsche held that passion could never be ruled by reason. It could be opposed and overcome only by another passion. In the politics of the plural soul, one passion (or set of passions) achieves a temporary victory after a struggle with other passions. Reason is subsequently employed— after the fact—to legitimate the results. Nietzsche was not wholly original in this assessment. In a famous passage from his *Treatise on Human Nature*, David Hume, the eighteenth-century Scottish philosopher, wrote, "Nothing can oppose or retard the impulse of passion, but a contrary impulse. . . . We speak not strictly and philosophically when we talk of the combat of passion and of reason. Reason is, and ought only to be the slave of the passions, and can never pretend to any other office than to serve and obey them." Hume went on to observe that some of our "calm" passions and instincts, such as benevolence, the love of life, and "the general appetite to good," are often mistaken for "the determinations of reason."[27] But in the end, only passions can motivate us to act in pursuit of any good or evil. The faculty of reason at best may be put to use to determine an effective means of achieving it.

Here Hume is equating reason with rationality. As such, reason is indeed a slave of the passions; it is an analytical tool put to use to achieve given ends. But when we encourage a person to "be reasonable," we are not asking that person to exercise logic in pursuit of any given end he might choose to pursue. Rather, we are urging the person to consider a wide range of goods, the physical and emotional impact of taking particular actions on oneself and others, and the benefits and costs of these potential outcomes. To ask someone to be rational is asking him to do just one thing. The complex activity of being reasonable, in contrast, goes far beyond an exercise in logic. It is a well-formed habit of exercising constitutional rule within a plural soul that inhabits an interdependent world.

Reasonableness is the disposition and practice of allowing the various passions or desires to engage in open debate, make their best cases, and, after all the votes are cast, follow through by putting policy into practice. John Dewey observed that "reasonableness is in fact a quality of an effective relationship among desires rather than a thing opposed to desire. It signifies the order, perspective, proportion which is achieved, during deliberation, out of a diversity of earlier incompatible preferences. . . . It is the attainment of a working harmony among diverse desires."[28] This working harmony among drives and desires is neither the despotic rule of a single passion (or group of passions) over reason nor the despotic rule of reason over the passions. Rather, it is the balanced coordination of the interdependent parts of the soul, a constitutional order within a confederation.

Following in the footsteps of pragmatist psychologist William James, John Dewey argued that passions gain power by keeping "imagination dwelling upon those objects which are congenial to it, which feed it, and which by feeding it intensify its force, until it crowds out all thought of other objects."[29] In this respect, the politics of the soul is a dangerous business. The angry person may whip himself into a frenzied state by imagining again and again, and with ever-greater intensity, the insults and injuries that originally provoked him. In similar fashion, the drug addict finds his desire for another "fix" crowding out every other desire—including the desire for his own health and welfare and concern for the well-being of family and friends.

While recognizing the dangers involved in the politics of a plural soul, Dewey wrote: "The conclusion is not that the emotional, passionate phase of action can be or should be eliminated on behalf of a bloodless reason. More 'passions,' not fewer, is the answer. To check the influence of hate there must be sympathy, while to rationalize sympathy there are needed emotions of curiosity, caution, respect for the freedom of others— dispositions which evoke objects which balance those called up by sympathy, and prevent its degeneration into maudlin sentiment and meddling interference."[30] In this respect, being "reasonable" is the good habit of imaginatively bringing passions and desires that may not be foremost into conversation with whatever impulses are currently in play. An inner debate of competing passions, with each ascending to the rostrum to make its case, is proposed. The merit of the outcome, Dewey argued, will depend on how well our diverse impulses get "interwoven."[31]

It is often said that the cure for the ills of democracy (of which there are many) is more democracy. Dewey gave the same recommendation with respect to the politics of the soul. The cure for the undeniable ills and dangers of an unruly passion is the consideration and interweaving of more passions. A constitutional regime characterized by psychic checks and balances is proposed.

Within this well-constituted *psyche*, reason is best understood as the capacity, and disposition, to employ imagination in our exercise of passion. Dewey writes: "We do not act *from* reasoning; but reasoning puts before us objects which are not directly or sensibly present, so that we then may react directly to these objects, with aversion, attraction, indifference or attachment, precisely as we would to the same objects if they were physically present."[32] Recall that our mirror neurons allow us to gain and refine skills simply by observing others (or even imagining ourselves) as they perform these skills. Likewise, we can respond emotionally to actions we might take and decisions we might make by playing out their performance and their potential outcomes in the mind's eye. The act of reasoning "puts before us" in imagination the emotional attachments and aversions that we would experience were a particular course of action taken. Effectively we emotionally live through narratives of possible futures. The fuller and deeper one's imaginative inhabiting of diverse narratives, the more reasonable one's decision making.

To render a reasonable judgment of an event, Adam Smith observed, we must "thoroughly enter into all the passions and motives which influenced it."[33] Making a reasonable judgment is not the exercise of rational calculation autocratically asserting itself over passions and desires. Rather, reasonableness is the capacity and disposition to entertain various passions and desires imaginatively—not only in their immediate gratification or thwarting but, as importantly, in scenarios where proposed actions have various effects and side effects. Empirical research confirms that emotion is, inevitably, a crucial component of reason.[34] If we fail to feel properly, we will fail to act reasonably.

In this respect, a rational decision is best seen as "an act of compromise" between the parts of the soul.[35] Importantly, many of the negotiations occur beneath the register of consciousness. In turn, the interweaving of competing desires and passions produces unpredictable results. This is disconcerting. Were the politics of the federated soul not characterized by uncertainty, however, psychological freedom would evade us.

Psychological freedom does not arise when a drive or instinct always gets its way, asserting its sovereignty. We do not consider an insect behaving in accord with its dominant instincts as enjoying freedom. And although the addict who is injecting himself with a drug is satisfying a sovereign impulse, we rightly consider him not a free man but a slave to his addiction. Psychological freedom is experienced not when an instinct rules supreme, but when the various parts of the human soul, notwithstanding potential conflict and competition, achieve a coordinated response. In the psychic community of the self, just as in the social community of the state, freedom can be experienced only within relationships of interdependence.

Individuation and Shadows

One of Freud's students, Carl Jung, addressed the challenge of turning the multiple soul into a confederated community.[36] He coined the term *individuation* to describe the challenge of integrating interdependent parts of the psyche into a functional whole that is expressive of freedom and capable of creativity.

Jung observes that the human ear can hear only a certain, limited range of sound frequencies (which is less than the range available to many other animals). Likewise, the human eye sees only a certain, limited range of light wavelengths. For example, we cannot see infrared light. Why then, Jung asks, should we be surprised that the human mind is aware of only a limited range of its own psychic activity?[37] Like Nietzsche and Freud, Jung held that what arises to our conscious awareness is but the tip of an iceberg. Most of what occurs in the psyche lies beneath the surface, within the unconscious.

Unlike Freud, Jung did not limit the unconscious to those parts of the psyche that have been repressed by the ego or superego. He describes the unconscious simply and inclusively: "everything of which I know, but of which I am not at the moment thinking; everything of which I was once conscious but have now forgotten; everything perceived by my senses, but not noted by my conscious mind; everything which, involuntarily and without paying attention to it, I feel, think, remember, want, and do; all the future things that are taking shape in me and will sometime come to consciousness: all this is the content of the unconscious."[38] There is no clean divide between the unconscious and the conscious. Things that are now known may later be forgotten. And things that we are unaware of, whether a deeply suppressed memory or the curious way we gesture while talking, may become the objects of exploration and investigation. That which is conscious, in any case, is never wholly conscious. There always remains some unknown yet significant component that falls below awareness.[39]

In addition to a personal unconscious, we also have a collective unconscious that, for Jung, includes all the "inborn" features of the mind that dispose us to think and act in particular ways. Jung observed that these tendencies are "handed down to us from primordial times in the specific form of mnemonic images or inherited in the anatomical structure of the brain."[40] The collective unconscious contains the "heritage of mankind's evolution, born anew in the brain structure of every individual."[41] Neuroscientists would refer to this heritage as innate neural mapping.

Our innate neural mapping produces instinctive ways of acting and intuitive ways of knowing. Jung labeled the inborn ways of knowing the world *archetypes*. "Just as his instincts compel man to a specifically human mode of existence," Jung writes, "so the archetypes force his ways of perception and apprehension into specifically human patterns."[42] Archetypes are uniquely human ways of perceiving and apprehending the world.

Action is made meaningful by placing it within a narrative that gives it a context and purpose. In the same fashion, Jung held, the world is made understandable by depicting it in archetypal images. Jung believed that mythology, along with dreams and art, provides access to the stock of archetypal images through which we perceive and apprehend the world. A myth is an account of events that never actually occurred but will always be true. It is a fictional story that symbolically captures the archetypal images structuring human perception and apprehension.

Jung's archetypes are often misinterpreted. They are not mystical, transcendent patterns, like Plato's heavenly Forms. Archetypes are simply products of the species' evolutionary history. They are specific neural maps that prompt us to see and understand the world in particular ways. Consider figure 6.1. Here you *perceive* two sets of three bumps atop and below one set of three bowls. What you actually *see* are six circles that are either light at the top, shading into gray in the middle, and dark at the bottom or, conversely, light on the bottom, shading into a gray middle band, and dark at the top. Why, then, do we perceive bumps and bowls rather than a bunch of variously shaded circles?

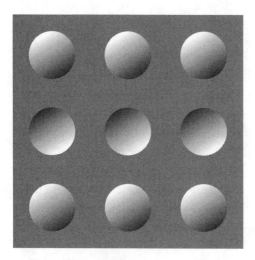

Figure 6.1
Bumps or Bowls?

The answer is that neural networks prompt us to perceive in three dimensions. Because we live in a world where the chief source of light, the sun, shines from above, these neural networks have evolved to interpret objects that have lighted tops and darkened bottoms as bulging out. Objects with shaded tops and lightened bottoms are perceived as concave. Jung speculated that archetypal patterns do more than structure our perception of physical space; they also structure our apprehension of social and psychic reality. We are mostly unaware of the patterns of perception that prompt us to interpret the physical three-dimensional world in particular ways. Similarly, we are mostly unaware of the archetypes that structure our interpretation of the social and psychic world. The goal of psychology, for Jung, is to gain access to these archetypes so we might become more aware of how and why the world means what it does to us.

Freud famously described the "Oedipal complex," wherein a male child instinctively lusts for his mother. A disciplining father figure who forbids the incestuous drive comes to be internalized by the child in the form of a superego. In this way, the child learns to restrain his own desire. But the child remains shamed by his erotic passions and guilt-ridden for his repressed impulse to kill the disciplining father. Jung also interrogated the myth of Oedipus. He agreed that desire for reunion with the mother was universal. But he argued that this Oedipal instinct was not a predisposition for incest. Rather, it constituted an impulse to become reimmersed in an all-protective and nurturing source.

In an uncertain and precarious world, rife with danger and disappointment, some part of us always wants to return to the safety of the womb, our earliest *oikos*. Jung found that the archetype of the "eternal mother" was helpful in providing meaning to the otherwise obscure behaviors of his patients when they pined for womb-like security in the face of life's challenges. Likewise, Jung explored the archetype of the hero, whose journey is celebrated in all mythological traditions. The hero represents the part in us that is willing to risk life, limb, or psychic security in order to achieve something great. The hero leaves the safety of the womb, the security of family, friends, and community, to explore and grapple with a larger, more dangerous world.

At one level, returning to the eternal mother represents a desire to live without the burden of consciousness, without the challenging struggles of a divided mind. For Jung, psychological health entails gaining access to the personal and collective unconscious without becoming wholly absorbed by it. Jung advocated the struggle of *individuation*. A form of self-realization, individuation is the product of a heroic grappling with instinctive and archetypal forces. It acknowledges their power without succumbing to them.[43] Individuation is a process of psychological development that values the growth of the conscious self's autonomy without losing sight of its roots in the unconscious. The individuating person pursues psychic wholeness by integrating rather than denying the soul's interdependent parts and hidden depths.[44]

For the newborn leaving the protective, fluid environment of the womb, the first breath of life is a painful affair. Until that point, the fetus and mother were, quite literally, one body. Separation causes suffering. The soul is a plurality in search of a lost unity. We cannot go back to the womb; we can never become whole again. But we have the option of becoming psychologically well integrated. We can achieve, through great effort, a working federation of the many parts of the soul. This is the task of individuation, or what Jung called "coming to selfhood."[45]

Coming to selfhood is a heroic journey. En route, again and again, the psychic adventurer must confront the archetype of the shadow—the part of ourselves that we do not integrate; it represents the feared, forbidden, or denied features of our unconscious. As the "ego-personality" of an individual is formed, aspects of his unconscious that prove too challenging to incorporate into his conscious sense of self get pushed aside. Unable or unwilling to accept these "dark aspects of the personality as present and real," the individual typically projects them onto others.[46] That is, he comes to see in others the dark aspects of himself that he refuses to integrate. Confronted with a world unconsciously made into "the replica of one's own unknown face," the individual becomes hypercritical of those features of others that he denies in himself.[47] Unwilling to confront the demons within, he projects them onto the world where he can battle them at a safer distance.

The shadow has dogged humankind throughout its journey of consciousness. Jesus Christ asked the hypocrites of his day why they perceived the speck in their neighbor's eye but did not see the log in their own. He counseled that they get rid of their own heavy timber before picking at another's sliver.[48] Christ was not simply an advocate of tolerance. He was addressing the archetype of the shadow. Those traits of others that particularly irritate us and produce self-righteous censure are unacknowledged and unaccepted parts of ourselves. We persecute in others that which we unwittingly condemn in ourselves.

John Dewey observed that "every impulse is, as far as it goes, force, urgency. It must either be used in some function, direct or sublimated, or be driven into a concealed, hidden activity."[49] Psychic energy, Dewey observed, is like physical energy: it can never be destroyed, only transformed. If our psychic energy is not acknowledged and used, it will come to "lead a surreptitious, subterranean life."[50] In the same vein, Jung argued that those parts of ourselves that we are loath to confront, our shadows, stand in particular need of being understood and their energies redirected. Otherwise they will wreak havoc on the world.

Acknowledging shadows is the first step to redirecting their force. The process of transforming unconscious psychic energy is neither simple nor easy, but a journey well begun is half done. To perceive the timber in your own eye is already to step away from hypocrisy and intolerance. Recognizing our shadows is half the battle; it marks the transformation of symptoms into symbols.[51] The self-righteous condemnation of others is the symptom of the unrecognized timber in the hypocrite's eye. When we display such symptoms, we literally do not know what we are doing. We misinterpret an effect for a cause. To identify the shadows that prompt our behavior is to transform murky symptoms into meaningful symbols. By recognizing these clandestine forces for what they are, we cease to be their victims.

To symbolize (from the Greek *sym bolein*) means to bring together. In naming our shadows, we symbolically bring together features of our unconscious selves and our conscious personalities. As Jungian theorist Edward Edinger observes,

To be able to recognize the archetype, to see the symbolic image behind the symptom, immediately transforms the experience. It may be just as painful, but now it has meaning. Instead of isolating the sufferer from his fellow humans, it unites him with them in a deeper rapport. Now he feels himself a participating partner in the collective human enterprise—the painful evolution of human consciousness—which began in the darkness of the primordial swamp and which will end we know not where.[52]

Until shadows are brought into symbolic light, they will do their dirty work in the dark. When we acknowledge them and bestow them with symbolic meaning, those parts of our unconscious selves that used to separate us from others (and perhaps prompt us to persecute others) now can reinforce a sense of community. In identifying a shadow, we recognize not only a dark part of ourselves but a force we have in common with all other human beings. That is the meaning of an archetype. Staying connected to your shadows is the best means of staying connected to others.

The unconscious is always active—we would neither breathe nor would our hearts beat otherwise. An inner politics is always in play, whether we realize it or not. Repressing shadows, instincts, and passions does not undermine their power. Quite the opposite. And it has a number of unintended consequences.

The denial or repression of shadows, instincts, and passions requires significant psychic energy. It is said that harboring a grudge is like keeping someone in prison. One has the sense of controlling the other person. But tremendous energy must be expended to "stand guard" at the door. The watchful grudge holder himself is in a kind of lockup. Effectively he has to stand guard over himself, ensuring that no fellow feelings or compassion escapes and finds its way toward the prisoner on the other side of the bars. Somehow he must rally the resources to ignore and contravene the work of evolution, which has fitted him with millions of mirror neurons. Likewise, keeping any part of the self suppressed demands significant effort. Psychic energy is a limited resource, so that which is expended suppressing part of the self is not available for more creative purposes. Moreover, the denial or repression of shadows, instincts, and passions is often counter-productive. Consider the challenge faced when you are given the command, "Do not think of pink elephants." The very effort to heed the command has you engaged in just what you are forbidden to do. In the same manner, the effort to suppress a part of the self typically gives that part greater power. This power generally exercises itself through projection.

The Neuroscience of Karma

In chapter 2, we observed that eating breakfast is doing more than one thing. In gaining morning nourishment, we are also supporting the corporations and political economies that produce the food on our tables. The act of eating is complex from an ethical point of view. It is also complex from a psychological point of view.

As newborns, we all eat in response to hunger pangs that prompt us to seek the nutrition required for health and growth. But most, if not all, of us also come to eat—at least some of the time—for other reasons. Certainly we often eat to sate hunger pangs. But we may also eat out of habit or emotional distress. We may chomp down finger food at a party, for example, to give ourselves something to do, to mimic our partners in conversation, or stave off social anxiety. We may eat particular types or amounts of food, such as chocolate or ice cream, to placate ourselves during troubling times or to satisfy a sweet tooth. We may eat as an addiction. Obesity is at record numbers across the world, and eating disorders plague contemporary societies. Heart disease is currently the leading cause of death in developed countries, and diet (along with lack of exercise) is the primary cause of heart disease. Clearly people find many reasons to eat beyond the demands of health and growth.

Ecologically speaking, no action has a single effect. Psychologically speaking, no action has a single cause. Our actions are products of complex negotiations going on within composite souls. To speak more simply: we typically act from mixed motives. We may be hungry *and* nervous when we make a beeline to the finger food table upon arriving at a social gathering. We may find a colleague's tendency to interrupt conversations a social deficit in need of correction. But the intense irritation it produces and the sharpness with which we rebuke him reflects as well our own unacknowledged struggle with impatience and self-absorption. Psychology informs us that the causes of action are as diverse as the parts of the soul, many of which reside in the unconscious.

Psychology also informs us that every action has multiple effects within the soul, many of them unintended. When we act owing to unconscious forces, in particular, we are not doing merely one thing. Pointing out the speck in someone's eye is not simply informing a neighbor of her faults. Neither is it simply getting something off one's chest. We are, at the same time, falling victim to an archetypal force. And in so doing, we reinforce its power, making similar actions in the future more likely.

Aristotle's habit theory of virtue helps us understand this phenomenon. Recall from chapter 2 that Aristotle claimed we acquire moral virtues by exercising them. To become virtuous requires doing virtuous things. Over time, doing virtuous things becomes a habit, a kind of second nature (with our instincts constituting first nature). Aristotle stipulates that "a man becomes just by the performance of just, and temperate by the performance of temperate, actions: nor is there the smallest likelihood of a man's becoming good by any other course of conduct."[53] Virtues are not good ideas that we embrace intellectually, endorse ethically, and subsequently decide to enact in our lives. Rather, they are habitual behaviors we come to adopt for various reasons, only one of which is the conscious, well-thought-out, articulate response that we give when asked to explain our actions. Of course, vice no less than virtue is a product of habit.

Physiologically speaking, you are what you eat. Psychologically speaking, you are what you do. Each of our actions reinforces the likelihood that we will take similar actions in the future. As Aristotle observed, "Habit . . . is practice long pursued, that at the last becomes the man himself."[54] Neuroscientists concur. Habits, including the complex habits we call virtues and vices, are neural maps that have become dominant through regular use. Just as the muscles in our bodies gain strength with use, so do neural maps. Neurons that fire together wire together. As a particular neural map in the brain is put to use, electrical impulses fire across its synapses (the junctions between neurons). The impulses strengthen the synaptic pathway, making it more likely that it will be employed in the future. Electrical impulses in the external world take the path of least resistance. The same is true of electrical impulses in the brain. The cerebral path of least resistance is a synaptic circuit well oiled from frequent use.

Nature is characterized by the survival of the fittest. Only those individuals and species best fitted to their environment endure over time. The neurological environment is characterized by the survival of the busiest. Synaptic pathways in the brain effectively compete against each other for electrical impulses. Those that keep busy by repeated use become dominant, which is to say, they become more likely to be used in the future.[55] Pathways that are not kept busy tend to atrophy. Effectively, a kind of neural Darwinism plays out in our heads. Behavior that is not reinforced by repetition loses the high-stakes neural game. An action, it follows, is not only a means to achieve a particular goal in the world. It is also, neurologically speaking, an end in itself. Behavior is self-reinforcing. At a neural level, behavior seeks to reproduce itself.

The notion of *karma* may be helpful here. The ancient Sanskrit word *karma* simply means action or deed. Within the Hindu and Buddhist traditions, the concept of karma explains how human action is related to the law of cause and effect. Each human action—including the activities of speaking, thinking, and willing—has effects. Because the world is portrayed within the Hindu and Buddhist traditions as a vast web of relations, the effects of actions are believed to find their way back to the doer. Like the wave produced by dropping a stone in the middle of a small pond, the effects of our actions ripple out in all directions, and, having reflected off the shore, find their way back to the center. Often karma is taken to mean that a personal scorecard is being kept. A tally of good and bad deeds is thought to yield an equivalent proportion of good and bad consequences for the doer. We hear people say, on witnessing a nasty bit of behavior, that it is "bad karma." Their expectation is that some form of retribution is forthcoming. Likewise, doing good things is thought to produce "good karma." Whatever goes around comes around.

At a neurological level, a scorecard is being kept, but it does not ensure reward or retribution for good and bad deeds. Of course, our actions do have effects on the world, and these effects on occasion return to help or hurt us. However, the most straightforward instance of karma occurs within the mind of the actor. Neurons that fire together (when we act) wire together (to make such actions more likely in the future). Karma is instantaneous. It all gets recorded at a neuronal level. There is no *away*.

Often we are unaware of the quality of mind that gives birth to a particular action. We do not perceive all our motivations. Acting in ignorance of unconscious desires and drives produces effects on the world and on ourselves. Jon Kabat-Zinn, a faculty member at the Massachusetts Medical School who employs Buddhist-style meditation to promote health, explains that

> every time we get angry we get better at being angry and reinforce the anger habit. . . . Every time we become self-absorbed, we get better at becoming self-absorbed and going unconscious. . . . Without awareness of anger or of self-absorption, or ennui, or any other mind state that can take us over when it arises, we reinforce those synaptic networks within the nervous system that underlie our conditioned behaviors and mindless habits, and from which it becomes increasingly difficult to disentangle ourselves, if we are even aware of what is happening at all.[56]

The ancient Law of Karma simply states that we become heirs to our own acts of will.[57] Whatever their worldly effects, our actions bequeath an inheritance within our brains in the form of reinforced neural networks. Every action leaves its legacy.

Neurologically, whatever goes around does indeed come around. Karma is concretized in the habits, dispositions, and tendencies produced by our thoughts, words, and deeds.[58] This neural repository may largely be sealed from the conscious mind. Importantly, the sealing off is itself a product of habituated action. Every action creates karma, which is to say that every action is habit forming. It makes a deposit, at a synaptic level, that earns daily interest. There is little, if any, forgiveness in the world of neurons and synapses. Everything leaves a trace, and everything matters. Whatever you sow, so you shall reap. As the Hua-yen Buddhists say, "He who kills another digs two graves."[59] The mind can never do just one thing.

Karma is not fate. Within the Buddhist and Hindu tradition, it refers to a dynamic relationship of cause and effect, not an unavoidable destination. Likewise, neural maps are not destiny. Although every action leaves its trace, these traces can be overwritten by new actions. Habits are notoriously hard to break, but with enough work, old habits can be replaced by new ones. Nature abhors a vacuum, and so does the psyche. "No psychic value can disappear," Jung said, "without being replaced by another of equivalent intensity."[60] Reshaping our neural networks is hard work, or, at least, long-enduring work. We cannot simply think our way out of bad habits. We must act our way out by forming and strengthening new habits of equivalent intensity over time.

The brain is a plastic organ that is shaped and molded by its environment and its own operations. Investigators of this neuroplasticity observe that "the experiences of our lives leave footprints in the sands of our brain like Friday's on Robinson Crusoe's island: physically real but impermanent, subject to vanishing with the next tide or to being overwritten by the next walk along the shore. Our habits, skills, and knowledge are expressions of something physical [in the brain's wiring]. . . . And because that physical foundation can change, so, too, we can acquire new habits, new skills, new knowledge."[61] In other words, the only way to get rid of bad karma is to replace it with good karma. Of course, thought and intention are also actions of a sort, and they too leave footprints in the brain. So the habit of becoming more mindful can be achieved only by practicing mindfulness.[62] In this respect, whether by thinking and intending, or taking concrete action, we are always reinforcing old habits or creating new ones.

To say that human beings are creatures of habit is not to endorse a conservative or reactionary ideology. One can develop a "habit of learning," even a habit of "inventive initiation." Habits can be progressive. If we are interested in changing the world, therefore, pitting creativity against habit is a mistake. If we want to promote creative actions and thoughts, we need innovative habits. As Dewey wrote: "To laud habit as conservative while praising thought as the main spring of progress is to take the surest course to making thought abstruse and irrelevant and progress a matter of accident and catastrophe. The concrete fact behind the current separation of body and mind, practice and theory, actualities and ideals, is precisely this separation of habit and thought." He argued that "those who wish a monopoly of social power find desirable the separation of habit and thought, action and soul, so characteristic of history. For the dualism enables them to do the thinking and planning, while others remain the docile, even if awkward, instruments of execution."[63] For Dewey, as for Aristotle, virtue is a habit. The only way to cultivate the virtues of inventiveness, openness, curiosity, and mindful awareness is to practice, again and again, these very activities.

Dewey held that the psychological development and well-being of the self was nothing more than the "working interaction of habits."[64] To integrate one's multiple habits and impulses into a coordinated whole is the achievement of character. Dewey wrote: "Character is the interpenetration of habits. . . . Integration is an achievement rather than a datum. A weak, unstable, vacillating character is one in which different habits alternate with one another rather than embody one another. The strength, solidity of a habit is not its own possession but is due to reinforcement by the force of other habits which it absorbs into itself."[65] For Dewey, having a strong character requires awareness of the constituent parts of the soul coupled with sustained effort to get the parts to work in concert. What the U.S. Constitution stipulated for the newly federated nation also applies to the composite soul. A "more perfect union" is required. But "a unity [gained] by oppression and suppression," Dewey observed, creates a house divided. It cannot stand.

How then do we know if a proper integration of the parts of the soul has been achieved? Inner resolutions that bespeak an ordered integration of habits and impulses are the product of and stimulate self-awareness. They leave us, as Kabat-Zinn suggests, in touch with the "quality of mind" that produced them. Importantly, this self-aware engagement in the politics of the soul is not a singular achievement but an ongoing practice. Just as the external world is ever changing, so the internal world of the psyche is fluid. As Dewey insists, "In quality, the good is never twice alike. It never copies itself. It is new every morning, fresh every evening. It is unique in every presentation. For it marks the resolution of a distinctive complication of competing habits and impulses which can never repeat itself."[66] Notwithstanding Dewey's earlier reference to the solidity of a habit, the ultimate goal of coordinating various tendencies and dispositions is to become adaptable. The well-integrated soul, while a creature of habit, forever acts in response to the unique confluence of events called the here and now.

With this in mind, attention to both the politics of the soul and the politics of the state becomes necessary, for the here and now to which we respond has its character significantly determined by the way we order our collective life. As political beings, we create the environment in which good and bad habits develop. We have the opportunity and responsibility to organize a political *oikos* within which the psychological *oikos* might optimally organize itself.

Loving Your Selves as Your Neighbors

The prefrontal cortex, the so-called executive brain, is primarily in charge of coordinating impulses and habits to achieve adaptive responses to changing environments. It marshals the neural networks of diverse areas of the brain into cooperative engagement. Elkhonon Goldberg observes that "the frontal lobes do not have the specific knowledge or expertise for all the necessary challenges facing the organism. What they have, however, is the ability to 'find' the areas of the brain in possession of this knowledge and expertise for any specific challenge, and to string them together in complex configurations according to the need."[67] In turn, the frontal lobes allow us to gain awareness of the various passions, desires, and habits that are operative in our actions.[68]

Importantly, the executive brain is not simply the home of reason. Emotion is required to motivate its operation and identify specific goals to be pursued. Cognitive neuroscience vindicates Hume and Nietzsche in these matters.[69] In the absence of the passion for psychological integration, the coordination of habits and impulses that allows the development of character and the making of good choices cannot get off the ground. But emotion is important in another respect. In order to integrate and coordinate our impulses and habits, we must become aware of them. This awareness is not a purely cognitive discovery. Rather, it is an empathetic connection. Once again mirror neurons appear to be involved.

Our capacity for self-reflection likely developed as a consequence of the work of mirror neurons in fostering empathy with the inner states of other people.[70] We first developed the capacity to imitate tribal members and read their hearts and minds to discern their intentions. This would have obvious social benefits and increase evolutionary fitness. Only subsequently did we turn these heart and mind reading skills on ourselves. The same networks of mirror neurons that originally were developed and used for the purposes of social cohesion and cooperation eventually allowed us to become aware of our own inner states. Interpersonal and intrapersonal attunement "share common neural correlates."[71] The neurological means we have developed to navigate social politics and soul politics are both grounded on empathetic connection. Our mirror neurons help us to know other people by allowing us to feel what they feel. They accomplish the same task in the same way for us.

Connecting to others and connecting to the self share the same neural correlates and present the same psychological challenge. Harry Haller, the protagonist of *Steppenwolf,* failed to gain attunement with either his own multiple soul or those who shared his world. Hess wrote that "his whole life was an example that love of one's neighbor is not possible without love of oneself, and that self-hate is really the same thing as sheer egoism, and in the long run breeds the same cruel isolation and despair."[72] Hess was inspired by Nietzsche, who also observed that if our souls are like "a knot of savage serpents that are seldom at peace among themselves," they will "seek prey in the world."[73] Effectively, Nietzsche was describing the psychological phenomenon of projection. The individual comes to see in others the dark aspects of himself that he is unable or unwilling to accept and integrate. And he makes them pay dearly for reflecting his shadow. The antidote for cruelty and other forms of prey seeking in the world, it follows, is for potential predators to become attuned to, embrace, and actively engage their own multiple selves and shadows. "We have cause to fear him who hates himself, for we shall be the victims of his wrath and his revenge," Nietzsche wrote. "Let us therefore see if we cannot seduce him into loving himself."[74]

Jesus Christ repeatedly endorsed the recommendation of Leviticus, the third book of the Hebrew bible: "Love your neighbor as yourself."[75] Here we are challenged to extend to others the same level of concern that we have for our own well-being. But there is more to it. Loving your neighbor and loving yourself are really two sides of the same coin. Indeed, Christ's edict might be reversed: the challenge is to love our multiple selves, the diverse parts of our soul, in the same manner that we love our worldly neighbors. This is not as easy as it sounds. There are parts of our souls that we know less well than we know our friends and fear more than we fear our enemies.

Perhaps the most famous accolade to the multiple soul embracing itself comes from Walt Whitman's "Song to Myself," found in his *Leaves of Grass*. Here Whitman famously stated, "I am large, I contain multitudes." Connecting to the multitude within and connecting to the multitude without are mutually reinforcing activities. Whitman wrote, "In all people I see myself, none more and not one a barley-corn less, And the good or bad I say of myself I say of them."[76] Psychological diversity and worldly diversity, for the American bard, require an equal embrace. Loving your neighbor as yourself becomes a feasible ideal with the realization that all my atoms as good belong to you.

Just as we all share the same atoms, so we all share the same components of the soul. It is possible to love your neighbor as yourself and to love your multiple selves as your neighbors only if you see in your neighbor the same inner plurality of which you yourself are composed—"none more and not one a barley-corn less." This is a prescription for empathetic understanding and psychological well-being.

It may also be a recipe for what psychologists call codependence, which occurs when there is an insufficient acknowledgment of boundaries. The codependent person gives up responsibility for his own life in a vain attempt to live vicariously. This may result in an effort thoroughly to control the other person or, alternatively, a willingness to be thoroughly controlled. Earlier we observed that the psychotic person lives in the isolation of an incommunicative independence. The codependent person falls into the other extreme: wholesale dependence. In both cases, the opportunity to embrace interdependence is forfeited.

The conscious development of interdependence is the achievement of a healthy psyche. It produces empathetic connection that respects another's individuality as well as one's own. Psychotherapist David Richo writes:

Adults learn that separateness is not an abandonment [of another person] but simply a human condition, the only condition from which a healthy relationship can grow. With boundaries comes interdependence rather than dependency. . . . To give up personal boundaries would mean abandoning ourselves! No relationship can thrive when one or both partners have forsaken the inner unique core of their own separate identity. . . . It is building a functional healthy ego to relate intimately to others with full and generous openness while your own wholeness still remains intact. This is adult interdependence.[77]

Achieving psychological interdependence entails loving one's selves— including the boundaries that allow the coordinated parts of the soul to become a unique individual—as much as one loves one's neighbors.

A growing field within psychology addresses the need to expand our neighborhoods beyond the human *oikos*. Practitioners of ecopsychology argue for and celebrate our fundamental connection to nature. Loving ourselves requires loving life. Psychological health is grounded in *biophilia,* the extension of our empathy and care to the organic world as a whole. Ecopsychologists maintain that that personal and planetary well-being are inextricably linked. Healing the mind and restoring the earth are complementary endeavors.[78]

Theodore Roszak, a proponent of ecopsychology, observes that this new field of inquiry explores the "greater ecological realities that surround the psyche." In turn, it denies that "the soul might be saved while the bio-sphere crumbles."[79] Indeed, the very dichotomy between self and environ-ment, between psyche and nature, is deemed unhealthy.[80] Bridging the separation of psyche and nature has been likened to gaining access to the unconscious and its archetypes. Ecopsychology fosters our caring for, and integration into, the natural world by way of the same processes that we care for and integrate the multiple components of our souls.[81] Our fear of nature's wildness, a fear that has been exploited to justify the domination of nature, often reflects a projected fear of our own shadows. To gain psychic and ecological health, much that we fear—the wildness of our instincts and the wildness of the natural world—must be accepted and embraced.

Nature is larger than us; that is one of its defining characteristics. The psyche is larger than the conscious self; that is one of its defining charac-teristics. Ecopsychologists maintain that the otherness of nature is a useful tonic for an ego that, as Jung said, "imagines he actually *is* only what he cares to know about himself."[82] As long as the ego thinks that "everything in the psyche is of its own making," the challenge of individuation cannot be met.[83] Confronting and celebrating the otherness of nature allows us to acknowledge and embrace the multiple parts of ourselves, the unruly oth-erness within that forever escapes our conscious awareness and control. As Rinda West observes, "This apprehension of a force that transcends will— the awe one feels beside the ocean, beneath the cathedral canopy of a great forest, or at the peak of a mountain—stimulates new ways of understand-ing oneself and the world."[84] From an ecopsychological perspective, the more we project our shadows onto nature, the more we will cut ourselves off from its potential to invigorate our lives.

Jung maintained that the collective unconscious establishes "the conti-nuity between the human psyche and the rest of organic nature." In theory, he argued, "it should be possible to 'peel' the collective uncon-scious, layer by layer, until we came to the psychology of the worm, and even of the amoeba."[85] Effectively, Jung was proposing that we all have a longbody. It stretches us across time and space, back billions of years through the evolution of life and across the entirety of organic world. It brings us into community with nature.

For the Iroquois and other aboriginal peoples, psychological health requires more than attention to the individual heart and mind, or even the hearts and minds of fellow tribal members. It requires the individual's full integration into the web of life. The aboriginal shaman or medicine person, David Abram observes, functions as an "intermediary between human and nonhuman worlds." Shamans, as ecopsychologists, derive the power to heal from their access and "allegiance" to an "earthly web of relations" in which the individual and her community are embedded. As Abram concludes, "We are human only in contact and conviviality with what is not human. Only in reciprocity with what is other do we begin to heal ourselves."[86] For ecopsychologists, truly loving yourself entails loving your neighbors, including those nonlegged, four-legged, six-legged, eight-legged, winged, finned, and leafy neighbors with which we share the world.

Conclusion

Nietzsche frequently invoked the ancient Greek poet, Pindar (c. 522–443 B.C.E.), who admonished his listener to "become who you are!" Psychologically speaking, realizing one's full potential is achieved by integrating the diverse components of the soul into a federation. One cannot integrate that of which one remains unaware. The first task therefore is self-knowledge.

To study the mind or soul in earnest is to acknowledge that its depths remain, and will likely forever remain, unplumbed. It is tempting, in the face of the inscrutability of the soul, to forgo any investigation. The ancient Greeks were aware of this temptation, so they took seriously the dictum inscribed on the portals of the oracle at Delphi: Know thyself! The Greeks would visit the oracle at Delphi to seek counsel. Her pronouncements were always enigmatic. In order to interpret well the oracle's words, self-knowledge was a prerequisite. Otherwise unrecognized desires, drives, fears, and projections might lead to dangerous misinterpretations. The best-known case is that of King Croesus of Lydia, who consulted the oracle to determine whether he should attack the Persians. Croesus was told that if he attacked, a great empire would fall. The Lydian king was elated by the oracle's response and quickly rallied his troops against the enemy. The Delphic pronouncement proved true, but it was Croesus's own empire that was destroyed.

Ambition may be blinding. If we remain unaware of our inner drives, we inevitably become their victims. There was, of course, another mythic king who unwittingly fulfilled prophetic words about his own downfall. As Jungian psychologist James Hollis observes: "Oedipus, who was the smartest man in Thebes, knew not himself, and that of which he was unconscious led to the fulfillment of the oracle's words." Hollis goes on to quote Jung: "When an inner situation is not made conscious, it happens outside as fate."[87] If we fail to recognize and wrestle with our shadows, they will have their way with us in the world. Ecopsychologists argue that the peril facing the planet is the direct product of our inner turmoil. The effort to safeguard the web of life, it follows, can be well pursued only in tandem with the effort to sustain our own souls. The converse also holds.

Shakespeare's *Hamlet* is another story of a prince who refuses to recognize his shadows. As a result, he condemns himself to inaction until a ghost forbids further denial of darker truths. At one point, Polonius, the father of Hamlet's beloved Ophelia, counsels his son, Laertes, to be true to himself. It is a worthy counsel. The self to which one must be true, however, is not a uniform and unchanging monolith that can be fully known and defined. It is multiple and dynamic. Being true to this active federation presents us with the challenge of integration. To have integrity is to be true to our own selves.

It is said that if you do not see your shadow, you are probably walking in the dark. To remain in the dark about our darker sides is a form of self-deception. Self-deception also occurs whenever we fail to realize that, psychologically speaking, we can never do just one thing. Every deed represents the final act of an inner political campaign, much of which took place below the register of consciousness. At the same time, every deed reinforces the strength of the victors of this inner campaign. It is self-deception, in this respect, to forget that we are the architects of our own souls. John Dewey observes that "self-deception originates in looking at an outcome in one direction only—as a satisfaction of what has gone before, ignoring the fact that what is attained is a state of habits which will continue in action and which will determine future results. Outcomes of desire are also the beginnings of new acts and hence are portentous."[88] The outcome of every inner political struggle is the foundation and framework for future campaigns. To become who you are demands the recognition that you are, with every action, setting the stage for the next act and for the further development of your character.

Not knowing oneself—psychological bystanding—is not doing nothing. It has effects on the self and on the world. As Nietzsche pointed out, when parts of the self are ignored or suppressed, they turn their wrath and revenge on the world, such that our neighbors become prey. Ecopsychologists argue that the domination of nature has a similar origin. Becoming more at home within our diverse psychic terrain allows us to become more at home in the world. Ecosophic awareness, for this reason, is inherently self-reflective.

Polonius's famous counsel to his son concludes with a tribute to the synergistic relationship between self and other:

This above all: to thine own self be true.
And it must follow, as the night the day,
Thou canst not then be false to any man.

Self-awareness is an empathetic relationship that we establish with our multiple souls. This loving inner attunement, from an evolutionary point of view and as a psychological phenomenon, is grounded in our capacity to empathize with our neighbors. At the same time, our empathy for other people, as psychologist Abraham Maslow stated, depends on the "inner integration" of our own souls. Integrated individuals— Maslow calls them "self-actualizing" people—are at home with the multiplicity within. Owing to their recognition and embrace of the wide range of impulses and passions of their own souls, they are able to connect with others without "fear of their own insides" getting in the way.[89] If the politics of the soul remains foreign to us, we will be unable, or unwilling, to grapple with the complex inner life that makes other people who they are.

To embrace the politics of the soul with all its interdependent relations—some of which will never fully break into consciousness—is to abandon the idea that one might ever fully master the self. At the same time, the effort is marked by dedication to self-knowledge, to the development of good habits, to the elusive work of integration, and to encountering openly a complex world. The soul or psyche is a place of relationships. Each of us is a multitude whose collective health depends on the dynamics of a sustainable integration. What is said here of individual souls applies equally to the *oikos* they inhabit. Sustaining our diverse psyches is crucial to the task of sustaining a diverse world.

7 Physics and Metaphysics

Nothing exists apart; everything has a share in everything.
—Anaxagoras

The earth, ancient peoples said, was a flat disc that supported itself in the ocean of space on the back of a giant elephant. The elephant, in turn, stood on the back of a giant tortoise. Asked what the giant tortoise stood on, one adherent of the archaic belief famously remarked: "Ah, yes, well after that, it's turtles all the way down."

The previous chapters have demonstrated that key fields of inquiry and facets of life are defined by relationships of interdependence. Examine one thing in these realms of study and life, and you find it connected to something else, which in turn finds support in yet another relationship. In this chapter, we examine concepts, theories, and discoveries in physical science and metaphysical philosophy that address the basic stuff of the universe—its most fundamental relationships. We explore the extent to which inter-dependence is built into the very fabric of the cosmos. In other words, we investigate whether it's turtles all the way down.

Physics addresses the elementary components of the universe and artic-ulates the laws that describe the movement of matter and the forms that energy can take. It pertains to the objective observation and mathematical modeling of natural causes and effects, describing the relation between material forces with such accuracy as to allow precise prediction. As a scientific study, physics produces knowledge by way of empirical experimentation.

Metaphysicians also inquire into the ultimate nature of reality, but they do not produce verifiable predictions. Through original thought and imagi-native speculation, they address deep ontological enigmas. Much of their subject matter concerns nonmaterial entities and relations and, hence, lies beyond empirical observation and confirmation. Eternal questions animate metaphysics.

Engineers, technicians, and other professionals regularly use physicists' discoveries to build and operate a vast range of tools and machines, including bridges, spaceships, lasers, computers, medical equipment, nuclear power plants, and bombs. Metaphysics does not have concrete applications. Much of the time, metaphysicians are preoccupied with questioning the meaning of their own enterprise. Unsurprisingly, physics and metaphysics are generally practiced by different people, in very different ways, and they seldom evaluate each others' work.

Why, then, write a chapter that couples physics with metaphysics? A brief account of the origins of these fields of inquiry provides an explanation. This historical exercise will also suggest why a book grappling with interdependence, unintended consequences, and sustainability concludes with a chapter that meshes physical science with philosophical speculation.

A (Final) Return to Greece

The first physicists were also metaphysicists. The so-called pre-Socratics of the ancient Greek world studied the heavens and the earth in an attempt to discern origins, establish natural foundations, and understand the mechanics of cause and effect. They were called *physiologoi* because they investigated *physis*, or nature, in the sense of the *nature of things*. They pursued rational explanations by invoking natural processes. Notwithstanding their claim to the discovery and practice of science—indeed, because of it—these early thinkers were also philosophers who speculated on the nature of the cosmos and the meaning of life. As "natural philosophers" they studied the workings of the parts, as scientists of our day do, but they also inquired into the nature of the whole.

Living from the seventh through fifth century B.C.E., the *physiologoi* were not satisfied with the answers to existential questions offered in mythology and religious tradition. Here the gods provide the explanation for all phenomena perceived but not understood. They turn night into day, throw thunderbolts, cause storms, and alternate the seasons. The pre-Socratics changed all that. Thales and his student Anaximander studied geometry, the weather, and planetary motion, predicting crop successes and solar eclipses, making celestial maps, and inventing sundials. Pythagoras and Parmenides studied mathematics and logic, discovering indubitable proofs, axioms, and laws grounded in the logic of noncontradiction. Empedocles, and later Democritus, speculated on the elements and mechanisms of basic physics, asserting that light travels with a finite speed, that air has a corporeal structure that exerts pressure, and that atoms—literally, things that are indivisible—constitute the fundamental building blocks of the universe.

Not infrequently, the pre-Socratics were persecuted, tried, and occasionally banished for their pursuit of natural explanations. Anaxagoras had the audacity to identify the sun and the moon not as heavenly gods but, respectively, as a large molten rock and a reflecting surface. He was thrown out of Athens. Xenophanes ridiculed the anthropomorphic polytheism of his day, suggesting that if oxen had hands with which to draw, they would depict all their gods as four-legged hoofed beasts. His idea was not well received.

For all their mathematical pursuits and scientific discoveries, the pre-Socratics were also philosophically minded. They invented what today we would call *meta*physics, literally that which is *beyond* physics.[1] They attempted to give meaning to life by understanding its significance within the most encompassing framework imaginable. They sought a *logos* of the *cosmos,* that is, an underlying rational foundation for the universe.

The pre-Socratics sought to discover an *arche*, a first cause or elementary principle established by empirical observation that would allow the development of a cosmology. To modern ears, these first principles seem idiosyncratic, vague, and utterly speculative. But for the first time in history, natural causes and effects, rather than theology and mythology, were employed to explain the origins and structure of the universe. Pythagoras, for instance, observed that different lengths of identical strings produced different sounds when plucked. Significantly, sounds that were harmonic to the human ear were produced when the lengths of strings were ratios of small whole numbers, such as 2 to 1, which produced an octave; 3 to 2, which produced a musical fifth; and 4 to 3, which produced a musical fourth. The physical world appeared to generate harmony out of arithmetic proportion. Discerning these mathematical foundations of the cosmos, Pythagoras proclaimed that "all things are number."

Thales, reflecting on the power of magnetism, discerned a basic energy that both the organic and inorganic world seemed to share. He speculated that a universal mind must infuse everything, linking all through an invisible power of attraction and repulsion. In the same vein, Anaximander claimed the *arche* was an unlimited, primordial entity called the *apeiron*, or indefinite. The *apeiron* was eternal, infinite, and perpetually generative of novel, distinct forms of matter. Heraclitus, observing the endlessly diverse and novel forms of life, said that "everything is flux." He conceived the universe as a unity of opposites. While nature often conceals the linkages that unite its diverse forms, one thing remains constant: change.

To say that the first scientists were also metaphysicians is not to slight their scientific investigations. Rather than developing replicable procedures to (dis)confirm empirical observations, as experimental scientists do today, the pre-Socratics employed a mix of observation, rational deduction and induction, and speculative imagination. In this respect, they were very much the forerunners of today's theoretical physicists, with the key difference being that the pre-Socratics (with the partial exception of Pythagoras) did not model the empirical world by way of complex mathematical formula. Rather, they relied on conceptual innovation to discern unity and harmony in a cosmos marked by diversity and endless change.

Physical cosmologists today, including astrophysicists, are carrying on the work that was begun by the pre-Socratics, superseded by luminaries of the "scientific revolution," such as Nicolaus Copernicus, Johannes Kepler, and Isaac Newton, and taken in radically new directions by Albert Einstein, who labeled his famous theory of general relativity a "cosmological consideration." The work of contemporary physical cosmologists is as scientific as any claim to knowledge. Yet their theories take us into a realm that radically challenges common sense. Indeed, contemporary cosmology and quantum physics are every bit as fantastic as the wildest assertions of metaphysicians. The difference is that modern physics is supported by strict methodologies and empirical validation.

People concerned with the sustainability of social and ecological systems have often echoed Wendell Berry's assertion that "the definitive relationships in the universe are . . . interdependent."[2] Is Berry's claim metaphysical in nature, affirming an unverifiable faith in cosmic unity? This chapter suggests that such claims have the weight of the most fundamental natural science behind them, and the backing of experimental proof.

Interdependence and Interpenetration

The story of Indra's net is meant to illustrate both the interdependence and the interpenetration of all phenomena. To this point, I have mostly addressed the phenomenon of interdependence, focusing on the link-ages—the physical connections between the strands. But the jewels hanging from the vortices of these filaments are not only physically tied to each other; they also mirror each other. Each jewel reflects not one but all the other jewels and, in turn, is reflected in the facets of every gem. Not only are all the parts connected; each part also contains the whole. This is the phenomenon that Hua-yen Buddhists called *shih shih wu-ai*, the interpen-etration of all things. Daisetz Suzuki defines interpenetration as "the One in the Many and the Many in the One."[3]

Martin Heidegger, who wrote extensively on metaphysics, noted that the perception of interpenetration was already well developed among the pre-Socratics. Heraclitus and Parmenides, Heidegger stated, "were still in harmony with the *Logos* that is, with the 'One is all.'"[4] *Logos* here refers to the unity of the cosmos. To be in harmony with this *Logos*, the diversity of life cannot be dissolved into a unified whole. Rather than overpower the many with the one, the pre-Socratics, along with some later philoso-phers, embraced the "harmony of multiplicity."[5] Interpenetration is not homogenization, the flattening of things to a uniform consistency. It bespeaks a multidimensional reality, a *multi*verse that is a *uni*verse.

Interpenetration might be understood as a deepened form of interde-pendence. It reflects an implicit or "implicate" order. The interdependen-cies of our world are difficult to uncover and often are ignored. But they are part of the explicit or "explicate" order. While interdependencies fre-quently go unheeded, no new physical laws or theories need be invented to account for them. A few decades ago, we did not know that the burning of fossil fuels in Nebraska was contributing to the melting of glaciers in the Himalayas and the rising of the seas in Polynesia. But these relation-ships are regulated by physical laws and remain available to empirical observation, investigation, and measurement. In this sense, they are explicit.

Interpenetration, in contrast, is unavailable to direct observation and defies commonsense understanding. Metaphysicians have speculated about it for millennia. Physical scientists began investigating interpenetration about a century ago. Only in recent decades, however, have they produced experimental evidence. These empirical demonstrations have not yielded a consensus within the scientific community as to the nature of interpenetration. But they are groundbreaking.

The phenomenon of interpenetration is often called *holism*. Sometimes holism refers to the study of complex systems from a multidisciplinary perspective. To study the world holistically, in this sense, is to approach it as an intricate web of relationships. In this chapter, holism is employed in a different sense. It refers to systems that display, at a minimum, the phenomenon of self-organization or *autopoeisis*. In turn, holism may refer to systems that display an implicate order. We might think of these two distinct phenomena as the holism of emergence and the holism of entanglement, respectively. But before we confront either of these forms of holism, we must tackle the physics of interdependence and the impact of unintended consequences.

Complexity, Chaos, and the Flap of Butterfly Wings

In 1952, the science-fiction writer Ray Bradbury wrote a short story for the magazine *Collier's Weekly* entitled "A Sound of Thunder." In it, a man from 2055 travels back in time to the days of the dinosaurs. He is under orders not to disturb the past in any way lest this disturbance have an impact on all subsequent history. Momentarily veering off a trail during his explorations, the time traveler unwittingly steps on a butterfly. When he returns to 2055, he detects subtle differences in how words are spelled, in the urban landscape, and, worst, in the rule of a fascist politician who, before the trip back in time was taken, had lost the election that would have placed him in power. The smallest change to an ecosystem millions of years ago produced crucial differences in the time traveler's present-day world. Bradbury was suggesting that everything is connected.

Twenty years after "A Sound of Thunder" was published, the mathematician and meteorologist Edward Lorenz gave a talk at the annual meeting of the American Association for the Advancement in Science. It was entitled, "Does the Flap of a Butterfly's Wings in Brazil Set Off a Tornado in Texas?" The answer that Lorenz gave at his 1972 talk was a resounding yes—at least potentially. For the previous two decades, Lorenz had been studying the nonlinear effects in atmospheric conditions that account for changing weather patterns. In the 1960s, his research had contributed to the creation of chaos theory. His now famous question about flapping butterfly wings graphically illustrated its major premise.

Chaos theory examines how small, imperceptible variations in initial conditions of dynamic systems, such as the weather, may produce large, observable changes in the behavior and structure of these systems. Lorenz's work, and chaos theory more generally, is often captured by the trope "the butterfly effect." The concept refers to the susceptibility of complex systems to significant disturbances owing to imperceptible and potentially distant events. Owing to the workings of nonlinear relationships, effects need not be proportional to causes. With multiple, minute causes capable of interacting to produce sizable effects distant in space and time, the overall system appears chaotic.

Chaos theory is really not about chaos. It does not refer to undifferentiated and wholly unordered states where things happen in a random fashion, as a product of chance. Rather, chaos theory applies to complex systems where order is not patent and events seemingly occur at random because their causes may be very small, very distant in time and space, and mediated by many other interactions and events. Henri Poincaré, a French mathematician and theoretical physicist, explored the foundations of chaos theory in the late 1800s. But it was the invention of computers more than half a century later that allowed Lorenz to fully develop it. Lorenz was running weather simulations on his digital computer and wanted to reexamine some data. To save time, he began his simulation in the middle of its course, employing the previous printout of a sequence of data. To his great surprise, the weather pattern that the computer spit out was markedly different from the previous simulation. Eventually Lorenz tracked down the problem. In the second simulation, he had rounded off a six-digit number from the first set of data to three digits (e.g., he might have entered 0.283197 as 0.283). Lorenz did not expect a difference at the level of a few parts in 10,000 in a single piece of data to significantly alter the results. But it did.

Until the development of chaos theory, Newton's laws of motion were thought fully to account for the mechanisms of our (macroscopic) world. Disputing Aristotle's belief that bodies tend toward states of rest, Newton insisted that a body in motion maintains its motion at a constant speed (velocity) until a force is applied to it, accelerating or decelerating its movement. Newton's second law, $F = ma$, is a linear equation that describes this relationship, where force (F) equals the mass of an object (m), multiplied by its acceleration (a), with acceleration understood as the rate of change in the velocity of an object over a set period of time. In crafting his laws of motion, Newton was able to explain both how an apple falls to the ground and how the earth travels around the sun, uniting for the first time the physical laws (of gravity) that rule the heavens and the earth. In Alexander Pope's immortal words, proposed for Isaac's epitaph, "Nature's laws lay hid at night; God said, let Newton be! And all was light."

If we know the height from which a bowling ball is dropped, we can use Newton's equation to predict its velocity when it reaches the ground. And we can know how long it will take to get there. The same predictions can be made if the bowling bowl rolls down a smooth, inclined plane, assuming we know the angle of the plane. We can also predict these values if we place the bowling ball halfway up the inclined plane or at any designated spot along its length.

Assume that at the top of a sharply inclined plane, another plane heads off in the opposite direction at a much shallower angle. Changing the spot from which we release the bowling ball on the steep plane changes the time it takes to hit bottom in a linear fashion. Move it up by a centimeter, and the time increases in direct proportion, as a square root of the change in (vertical) distance. But when we move the bowling ball up the last centimeter, to the cusp of the apex, a mere nudge would send it down the more gradually inclined ramp. A small change in the initial condition at this point dramatically alters the time the bowling ball takes to reach bottom, as well as its terminal velocity. Here, at the apex, a minute shift in the initial condition produces very different results.

Complex systems might be thought of as places filled with many such apexes. It is not that the laws of physics do not apply in these systems. They do, just as Newton's laws of motion apply regardless of which inclined plane the bowling ball descends. But the events of such systems remain very difficult, if not impossible, to predict. One might think of complex systems as inclined planes that are not uniform, flat surfaces, but bumpy hills covered with stones and boulders, plants and trees, nooks and crevices. These irregularities produce multiple apex-like interactions. Calculating the time a ball takes to roll down this bumpy, obstacle-strewn surface, the terminal velocity it will reach, the path it will take, and where it will land becomes extraordinarily difficult. Release any two balls from the same spot at the top of such a hill, and the chances of their following the same path in the same time and ending up at the same spot at the bottom are very small.

Consider the fate of some itinerant rubber ducks. In 1992, a cargo ship plying the waters of the Pacific lost twelve containers overboard in a storm. One was filled with 28,800 "Friendly Floaties"—rubber ducks and other animal toys designed for toddlers' bathtubs. The container broke apart in the water, and its plastic pilgrims began their journeys. Ten months later, the first of the floating toys came ashore in Sitka, Alaska, over 2,000 miles away. Some took years more to make the same journey, and others visited various other ports of call around the world.

At one point, all the bathtub toys were, quite literally, in the same boat. When the container broke apart in the open seas, one might imagine all 28,800 plastic ducks, turtles, beavers, and frogs to be in the same initial condition, located within clear sight of each other in the same geographical location, ocean current, and weather pattern. Actually, their initial conditions were slightly different. So a wave here and a breeze there sent the toys on distinct journeys. In the ensuing years, the Friendly Floaties have found their ways to both coasts of North America and to three other continents. Many remain at sea today, with at least one flotilla predicted by oceanographers to make landfall in Britain in the near future.[6]

The world presents us with complex, chaotic systems, like the atmosphere and the ocean; very simple, predictable systems, like the swing of a pendulum; and everything in between. Likewise, every field of inquiry, be it biology (from the structure of cell bodies to random mutation) or mathematics (from the straightforward axioms of Euclidean geometry to Kurt Gödel's Incompleteness theorems, which demonstrate that no system of axioms can include all proofs or prove its own consistency), addresses both simple and complex systems.[7] The difference between predictable and chaotic systems is primarily the number and variability of the relations of interdependence that compose them. An object with relatively few forces acting on it, like a bullet shot out of a gun, follows a highly predictable path. An entity made up of countless interactive components and processes, like a hurricane, does not follow a highly predictable path. Still, the trajectory of the bullet and the path of the hurricane are both part of the explicate order. Complex systems are highly variable and effectively unpredictable because small, often imperceptible interactions among the parts produce large disturbances over time. But the forces that produce these interactions are determinate and potentially knowable, at least at the macroscopic level.

That, at least, was the position of French mathematician and astronomer Pierre-Simon Laplace (1749–1827), occasionally referred to as the "French Newton." In his *Philosophical Essay on Probabilities*, Laplace wrote:

We may regard the present state of the universe as the effect of its past and the cause of its future. An intellect which at a certain moment would know all forces that set nature in motion, and all positions of all items of which nature is composed, if this intellect were also vast enough to submit these data to analysis, it would embrace in a single formula the movements of the greatest bodies of the universe and those of the tiniest atom; for such an intellect nothing would be uncertain and the future just like the past would be present before its eyes.[8]

For Laplace, the solution to the problem of prediction was simple: get more data. The universe is not fundamentally chaotic, just complex. To be sure we remain ignorant of the precise positions, velocities, and masses of all the bodies that compose it. Were this information available to us, however, we might apply Newton's laws to predict the future with 100 percent accuracy. The universe is not indeterminate, according to Laplace. It is merely intricate.

Quantum Physics and Fundamental Uncertainty

Since the advent of quantum physics, Laplacean faith in the power of information to produce accurate predictions has been shattered. It is not that Newton's laws, or cause-and-effect relationships more generally, no longer apply. The universe is not wholly chaotic. Indeed, it continues to operate in a very ordered and predictable manner. Certainly at the macroscopic level—from the swirling of galaxies, to the rotation of planets, to the trajectories of missiles, and even to the motion of molecules—the universe is a law-abiding place. That is why engineers and technicians can achieve amazing feats by applying the laws of physics and why physical scientists can explain and predict so much of our world. But if we investigate the universe at the subatomic level, these mechanistic regularities break down. Here new laws, those of quantum physics, must be applied.

The laws of quantum physics do not speak in the language of cause-and-effect certainties, as Newton's laws do. Rather, quantum physics employs the language of probability. Uncertainty and chance play significant roles within this lexicon. The quantum world exists as a matrix of possibilities. The potential states of this world gain determinate value only through their measurement. The famous uncertainty principle developed by Werner Heisenberg (1901–1976), the German Nobel Prize–winning physicist, reflects this uncanny feature of the quantum world. Developed in the late 1920s, the uncertainty principle constitutes something of a watershed event. It distinguishes quantum physics from classical mechanics.

Perhaps the best way to understand Heisenberg's uncertainty principle is to return momentarily to Laplace. Chaos theory asserts that small changes in initial values (e.g., the flapping of a butterfly's wings in Brazil) can make big differences in end states (the formation of tornadoes in Texas). We cannot predict the end state because we cannot know every initial value with sufficient certainty. Recall that Lorenz discovered this "butterfly effect" because he shortened a particular initial value by a few decimal places to save time when entering data into the computer. Of course, even the original, unabbreviated initial value would not have been wholly accurate. It was precise only to six decimal places. There is nothing magical about the number six: seven or eight decimal places would have been even better. In Lorenz's atmospheric studies, as with any other effort to understand a complex system, levels in accuracy of initial values reflect the limits of our instruments of measurement. This is true even for the simplest systems. Billiard balls banging about a table act predictably. But they cannot be weighed with infinite accuracy; neither can the angle of incidence of their collisions be exactly measured. It follows that their interactions and final configuration can be predicted only within some limited degree of certainty. To predict the behavior of any system with complete accuracy, one would need initial values measured with infinite precision.

Infinite precision is obviously impossible to attain in practice. Heisenberg demonstrated that it was also impossible to attain in theory. Heisenberg's basic principle is simple enough: the more accurately one determines the position of a subatomic particle such as an electron, the less accurately can one determine its momentum (i.e., its mass times its velocity).

Contrariwise, the more precisely one measures its momentum, the less one can know about its position. In other words, all initial values cannot be determined with accuracy at the same time. Hence the trajectory of objects, and their subsequent interactions, can no longer be predicted with precision.

Heisenberg explained his principle with the illustration of an imaginary microscope. When we measure the weight, size, and velocity of something, we typically observe, respectively, its impact on a scale, its size in reference to a ruler or metric, and its speed in reference to a metric and clock. In each case, these measurements become visible to us owing to the light that bounces off the object, and off the scale and metric, into our eyes. To determine the position and momentum of a subatomic particle such as an electron, we need to bounce light off it and observe the results. Effectively, one has to shoot an even smaller particle, a photon of light, at the particle one is attempting to measure and observe what this photon does.

Light, and electromagnetic radiation more generally, comes in particular frequencies or wavelengths. The position of an electron can be measured accurately by using a photon of high-frequency light, say, gamma rays. One might think of high-frequency light as a ruler with very small spaces between its markings. However, photons with a high frequency also have high energy, so when gamma rays strike an electron with lots of energy, the electron recoils. Effectively the act of measuring its position changes its momentum.

Alternatively, if the photon used for measurement has a lower frequency, like visible light, the momentum of the electron can be measured quite accurately because the photon delivers less energy to disrupt it. But then the electron's position cannot well be determined because the frequency of the light is too low to pin down the location of the electron with precision. Measuring the position of subatomic particles with low-frequency light is like trying to determine the exact location of a bumblebee when the only information one has is the time taken to complete its trip from the hive to a flower patch and back. Although the bumblebee's average speed can be easily calculated (by dividing twice the distance between hive and flowers by the elapsed time), one would never know the precise location of the zigzagging insect at any particular moment.

When it comes to measuring the initial values of subatomic particles, Heisenberg discovered, you can never do just one thing. Determine the position of a particle with great accuracy, and you affect its momentum.

Measure its momentum with great accuracy, and you blur its position. Measurement itself has unintended consequences. The event (of measuring) changes the entity (being measured). Indeed, subatomic particles such as photons and electrons may be better understood as events themselves rather than things. Measuring them becomes part of the event they are. As Heisenberg wrote, "What we observe is not nature itself, but nature exposed to our method of questioning."[9] At the subatomic level, in the realm of quantum mechanics, measurement is an act of intervention—the onset of a relationship of interdependence between the observer and the observed.

Quantum physics is grounded on the fact that the position and momentum of an elementary particle—and, for that matter, its very existence—are subject to uncertainty. At least to human observers, the entity exists only as a probability. At the quantum level, physicists cannot say categorically if a particle exists, where it is going, or how fast it is moving. They can speak of only a range of probability that such events are occurring.

We often calculate general probabilities without being able to say anything with certainty about individual events. The fields of statistics and actuarial science were specifically developed to address just such aggregate trends. Actuaries employ statistical methods and mathematical modeling to assess risk and probability for financial industries. By studying aggregate trends regarding life expectancy and causes of death, for instance, actuaries help determine payment rates for (classes of) clients of life insurance agencies. If they get their numbers wrong, the insurance business may consistently lose money. So the science is quite well developed. Actuaries cannot say with certainty, or even great probability, how or when any particular client will leave this world. But they can predict, with reasonable accuracy, the percentage of clients of various age groups who will still be alive in ten years' time.

Quantum physicists employ statistical sciences in this sense. They cannot say with certainty what any particular subatomic particle will do, but they can speak with authority about the aggregate results of many such interacting particles. Still, there is a sense in which the uncertainty involved in quantum physics goes beyond the uncertainties that actuaries face. For quantum physicists, uncertainty is not only a product of epistemology (how our knowledge is gained); it is also the product of ontology (the fundamental nature of reality). As physicist David Bohm writes,

"Heisenberg's principle should not be regarded as *primarily* an external relation, expressing the impossibility of making measurements of unlimited precision in the quantum domain. Rather, it should be regarded as basically an expression of the incomplete degree of *self-determination* characteristic of all entities that can be defined in the quantum-mechanical level."[10] The problem is not simply that we are uncertain in our measurements. Subatomic particles themselves embody uncertainty. To be sure, part of the problem is that we can never do just one thing. Observation changes reality. But the more fundamental problem, one might say, is that elementary particles are not just one thing. The wave-particle duality of light illustrates this puzzle.

Light behaves like a wave. Specifically, it demonstrates the property of dispersion, splitting up into varying frequencies or wavelengths. That is why (white) light traveling through a prism produces a rainbow of color, with each color representing light of a different wavelength. Light also demonstrates the property of diffraction: it curves around obstacles. Waves of light (like waves of water) that travel through a narrow channel will spread out in concentric circles on the other side, effectively bending around the corners. Sunlight curving around the moon during a solar eclipse illustrates the same phenomenon. And like a wave, light demonstrates the property of interference, producing particular patterns when two sources of light cross paths. Just as two intersecting waves of water produce extra high crests wherever two crests meet, and extra deep troughs wherever two troughs meet, so waves of light crossing paths create patterns of very light and very dark bands. Finally, like a wave, light demonstrates the property of refraction, changing direction and speed whenever it enters a new medium. That is why a straw in a glass of water appears bent: the waves of light traveling between the straw and one's eyes change direction when they move between water and air.

Light also behaves like a particle. Indeed, a photon of light might be thought of as a high-speed particle of no mass. Like any other projected object, a photon of light can travel through a complete vacuum. Other waves, such as sound waves or ocean waves, require media through which to pass. That is why a bell rung in the vacuum of space makes no noise. And like other particles, photons of light travel in a straight line until another force changes their direction.

Physicists coined the term *wave-particle duality* to convey the fact that light sometimes acts as if it were a particle and sometimes as if it were a wave. Occasionally the term *wavicle* is used to describe photons of light (or other elementary particles) that demonstrate this duality. As a particle, a photon of light might be thought of as having a particular location and momentum at any particular time, even if the attempt to measure these values precisely proves impossible. But as a wave, a photon of light exists only as a range of possible locations and momenta. Light itself, the means by which we measure other subatomic particles, has uncertainty built into its very structure.

One might intuitively understand this phenomenon by comparing projectiles to waves in the macroscopic world. It is obvious to those who are playing catch where the ball begins and ends. The ball is discontinuous with its environment, that is, its boundaries are clear and distinct. No one is going to confuse the ball (the projected object) with the air (the medium) through which it passes. Indeed, the ball would travel just as well (indeed, even faster and straighter) were there no air present. Now consider a wave such as that sent back and forth along a skip rope that is undulated between two children. Can we distinguish the wave from the medium through which it passes? The wave could not exist without the rope, but the wave is not the rope itself (for the rope could be taut and still). Unlike a ball traveling through the air, a wave traveling through a rope is not discontinuous with its medium. Its boundaries vary. Or think of a slipknot tied in a rope: the knot can be moved up or down the rope without changing the rope's structure. The knot is not the rope itself, but only a spatial arrangement of it. Although the boundaries of the knot change continuously as it is slid up and down the rope, it maintains throughout what Buckminster Fuller called its "pattern integrity." In the same way, a wave is a spatial displacement of a medium that maintains pattern integrity over time.

Waves of light are energetic (rather than spatial) displacements. They have no determinate boundaries but rather exist as a range of probability. This fact was demonstrated by the famous double-slit experiment. If photons of light are shot through a tiny slit, a fanned-out distribution will appear on a screen placed on the other side. The screen will be brightly illuminated directly opposite the slit and increasingly darker toward the periphery. One might conclude that most photons pass through the slit in a straight line, with ever-smaller numbers coming at angles to produce the diffraction pattern. They act like the hits of a good marksman, with most in the middle of a target and progressively fewer distributed around its periphery, with occasional strays missing the target altogether.

When photons are shot through two parallel slits, one might expect the diffraction patterns on the screen simply to overlap, much like the effect produced by two marksmen standing side-by-side taking aim at targets placed next to each other. In fact, this does not occur. Rather, an interference pattern is produced on the screen, with a series of light bands alternating with dark bands. This interference pattern makes no sense if one understands photons to be particle projectiles moving according to the laws of classical mechanics. However, if one understands photons to be waves of electromagnetic energy, then an interference pattern of highly illuminated crests interspersed with deeply shadowed troughs is exactly what one would expect. So the double-slit experiment confirms the wavelike nature of light. Indeed, when Thomas Young first carried out the experiment in 1803, he single-handedly destroyed Newton's then dominant corpuscular (particle) theory of light.

The wavelike character of light is curious. But things get even stranger when a single photon of light (or a single electron) is shot at a double slit. With a single photon passing through one or the other slit and hitting the screen on the other side, one might assume that an interference pattern could not form. Presumably the single photon would never physically encounter any other photons, so there could be no interference pattern. But in experiments employing single photons (or electrons) shot one at a time at double slits, an interference pattern does indeed form on the screen. Close one or the other slit, and a typical diffraction pattern appears. Open both slits, and light and dark bands appear. When photons are shot one at a time at double slits, the screen effectively displays the potentiality of their interference. It is as if the single particle passing through one slit acknowledges the possibility of its having passed through the other slit. Accordingly it displays this potentiality on the screen by creating an interference pattern.[11]

The double-slit experiment has been carried out with electrons since the 1950s, neutrons since the 1970s, and atoms since the 1980s.[12] It demonstrates the fundamental indeterminacy of quantum events. The interference patterns formed in these experiments indicate that no matter how great the accuracy our tools of measurement achieve, we can never say with certainty where a single particle will strike. All we can observe (and measure) are realms of probability. Physicists chart these potentialities mathematically as wave functions or probability waves. The crests and troughs of these wave functions signify, respectively, high and low levels of probability that the events in question will occur.

It appears that the probabilistic, indeterminate nature of the quantum world is not simply a product of our limited means of gathering information. Here the continuity (and hence determinacy) of motion and the mechanisms of cause and effect—as we have come to experience these phenomena in our daily affairs and accurately measure them since Newton—no longer hold sway. In the previous section, we learned why small changes in initial values in a complex, chaotic system may have large consequences, making accurate prediction impossible. Now we have learned that at the level of elementary particles, this unpredictability is not simply the product of our inadequate instrumentation and poor means of measurement. When the initial values in question occur at the subatomic level, chance and indeterminism play an essential role.

The Holism of Emergence

In the early 1920s, Lloyd Morgan published *Emergent Evolution,* marking the origin of an idea that has been debated ever since.[13] It was only in the 1960s, however, that the fields of ecology and quantum physics vibrantly revived the debate, primarily through the work of physical chemist Ilya Prigogine. Emergence refers to the arising of a property in a whole that does not appear in its separate parts. The properties of water, for example— its liquidity at room temperature, its freezing point, boiling point, and chemical interactivities—are not to be found in, and could not be deduced from, the properties of hydrogen and oxygen, the two elements composing H_2O. Likewise, the temperature of a gas is not to be found in, and cannot be deduced from, the characteristics of a single gas particle. In a gas, temperature refers to the average kinetic energy of its particles. The faster the gas particles are moving relative to each other, the hotter the gas. So it does not even make sense to speak of the temperature of a single gas particle (regardless of how fast it is moving). You need interaction—molecules bumping into each other—to get heat. Temperature is not a characteristic of the isolated parts of the gas. It is found, and can be measured, only in the gas as a whole.[14]

Emergence typically refers not only to the arising of a property in the whole that does not exist in a single part (e.g., the temperature of a gas) but the arising of a property in the whole that does not exist even when all the parts are aggregated. The classical statement of this sort of emergence occurs in Aristotle's *Metaphysics*. Aristotle was investigating what makes a bunch of things into a unity. What is it that emerges, he asked, when a heap becomes a whole?[15] In asking this profound question, Aristotle was inquiring about a particular whole—the human being.

Our bodies contain sixty chemical elements, with over 98 percent of their mass made up of just six elements: oxygen, carbon, hydrogen, nitrogen, calcium, and phosphorus. What would happen if you measured out exact proportions of all sixty elements, put them in a large container, and shook it vigorously? The answer is, "Nothing much at all, apart from a soupy mess." Certainly human life would not be created. Regardless of how well one understood the individual properties of these sixty chemical elements, one would know next to nothing about the entity we call a human being. The same is true when the whole in question is a single cell rather than an entire organism. The cell, no less than an animal or plant, displays characteristics that cannot be deduced from the properties of its constituent (chemical) parts. In each case, life is an emergent property of certain complex systems of molecules. To understand life, we must study its diverse forms as wholes. Studying the parts (elements) in isolation does not get us very far.

Nobel laureate physicist P. W. Anderson once observed that "more is different."[16] The whole is not simply an aggregate of parts. When enough parts are added to certain physical systems, more is not simply more (of the same); it becomes different. As the quantity of components grows, so do interactions among them. By means of these interactions, quantitative differences get translated into qualitative differences. *More* eventually becomes *more complex*, not simply *more abundant*. With more components, more time, more interactions, even more laws and rules, we witness the emergence of new properties.[17] This is obviously true when aggregates of molecules combine to form living organisms. But even at the level of particle physics, Anderson insists, we may observe the emergence of new properties that cannot be understood as the simple extrapolation of the properties of the constituent parts.[18] Anderson's insight is often encapsulated by the expression, sometimes attributed to Aristotle, that "the whole is more than the sum of its parts." This is decidedly the case, as we have seen, when the whole in question is a human being rather than a heap of chemicals. More is also different when the whole in question is a human organization. Political bodies no less than physical bodies display emergent properties that their constituent parts lack. Focusing on relationships, not elements, proves key. This is the central insight of systems thinking.[19]

The ecological, ethical, technological, economic, political, and psychological realms of our world cannot well be understood by employing a mechanistic model. That is because the parts of these complex systems exist in networks of mutually conditioning relationships that together display emergent behavior. In such situations, emergence is defined chiefly by the qualities of novelty (i.e., new properties), adaptability, synergism, nonlinearity, self-organization, and history. History, in the sense of developments that heed the arrow of time to yield particular, irreversible sequences of events, is particularly important. You cannot understand the meaning of a sentence, paragraph, or book simply by understanding the meaning of all the words it contains. Placed in a different order, the words of a sentence might mean something quite different. Similarly, complex systems display their properties owing to the development of events and relationships in a particular order. Even when we understand all the laws governing each of the parts, these laws cannot be employed to (re)construct the whole.[20] We need to know something about the sequence of events. History is crucial.[21]

Reductionists like Laplace posit the whole simply as the product of a chain of antecedent, mechanistic relationships. They insist that the whole is nothing but the sum of its parts. But reductionism can be made compatible with the holism of emergence if we define *sum* and *parts* in a particular way. If by *parts*, we mean the most basic elements of a system, and if by *sum*, we mean a simple aggregating of interactions and properties, then systems displaying emergence are indeed more than the sum of their parts. However, if by *parts*, we mean not only the basic elements of a system but also the relationships that these elements form with each other, and if by *sum*, we mean not only the interactions and properties of parts but also the synergistic effects of these interactions and properties as well as the particular sequencing of their interactions in time, then the whole can indeed be reduced to the sum of its parts. An expansive reductionism of this sort is compatible with emergence. That is why a complex system displaying emergence is not wholly unpredictable.

When hydrogen and oxygen are combined under certain conditions, for instance, we can predict with great certainty that water will form and what properties the water will display. We can also predict with great certainty the behavior of more complex systems and organisms, such as cells, plants, animals, and certain social structures. This is precisely what the natural and social sciences do. Of course, the behavior of highly complex systems is not predictable with 100 percent accuracy. To the extent that quantum events affect initial values, mechanistic explanation (Laplacean reductionism) breaks down. But quantum uncertainty is not the only threat to predictability. In Laplace's universe, parts can interact, quite literally, only by bumping into each other. Only direct contact with another object causes positions and velocities to change. All physics, one might say, is local. This truth applies well enough in Newton's mechanistic world. In the quantum world, however, physics does not restrict itself to local effects.

The Holism of Entanglement

We take gravity for granted. That is sensible. Ignore the law of gravity, and the results are seldom good. Yet gravity remains an enigmatic force that defies understanding for even the most advanced physicists. We still do not have an adequate grasp of how it works or even what it is. Gravity proves such a conundrum in part because it appears to exhibit the phenomenon of action at a distance. To have action at a distance, one must have two bodies separated by space that demonstrate a causal effect of one on the other. This effect must occur faster than any information might be conveyed between the bodies, thus ensuring that the effect is not simply the product of unseen (mechanical) impacts. Gravity appears to produce just such an effect.

Gravity can affect the direction or momentum of distant objects, even across the vacuum of space. Although Isaac Newton developed the formulas that allowed calculation of the force of the earth's gravity, he could never quite accept the notion of action at a distance. Newton wrote, "That one body may act upon another at a distance through a vacuum without the mediation of anything else, by and through which their action and force may be conveyed from one another, is to me so great an absurdity that, I believe, no man who has in philosophic matters a competent faculty of thinking could ever fall into it."[22] Of course, gravity seemed just such a force that acted at a distance without mediation across a vacuum. It was a force Newton could measure but not understand.

Still grappling with the problem of gravity centuries later, Albert Einstein developed his general theory of relativity. The capacity of gravity to act at a distance was ingeniously solved by proposing that matter warps or curves space and time. Consider a large body such as our sun with an orbiting satellite, such as the earth. If the sun was suddenly obliterated at time x, Einstein asked, would the earth cease to feel the effects of gravity at time x? If the answer is yes, then the force of gravity, demonstrating an immediate effect, would have moved faster than the speed of light. Yet nothing can move faster than the speed of light, which is a universal constant. If the answer is that the gravitational pull on the earth ceases very quickly but not immediately on the obliteration of the sun, then you have a very strange situation. For a span of time, the earth would continue its circular orbit around a nonexistent entity.

Einstein solved this paradox by supposing that a massive body like the sun curves space and time around itself. This curvature, caused by gravitational waves, travels at the speed of light. A planet orbiting the sun is effectively always falling into this curvature, much like a ball lying on the edge of a trampoline will always roll in toward a person standing in the center. When the sun is obliterated, the gravitational wave that causes the curvature of space and time would cease. So the earth would stop being affected by the curvature about eight minutes after the sun was obliterated, which is the time it would take the gravitational wave (and the last rays of light) from the just obliterated sun to reach the earth, some 92 million miles away. Again, one might visualize the ball along the edge of the trampoline continuing to roll inward for a brief period after the person standing in the center is vaporized because the original curvature of the elastic trampoline would require time to flatten.

Einstein demonstrated that space and time, formerly assumed to be distinct and unrelated, were in fact interdependent phenomena. Indeed, they were so closely related that he coined the term *spacetime*. Einstein also demonstrated in theory, as the scientists who split the atom demonstrated in practice, that mass and energy (the potentiality of action) are not distinct and separate entities. One can be transformed into the other. That is the meaning of Einstein's famous equation $E = mc^2$, where the energy of a body at rest is equal to its mass times the square of the speed of light. Mass is a form of congealed energy, just as light, radiation, and heat are other forms of energy. Space, time, mass, and energy participate in relations of interdependence.

Einstein, with two colleagues, Boris Podolsky and Nathan Rosen, hypothesized an ingenious experiment in 1935, now known simply by their initials: EPR.[23] The hypothetical experiment explored the notion of action at a distance, which physicists now call *nonlocality*. EPR, originally an exercise in theory, has been empirically confirmed numerous times over the past half-century.[24] In an EPR experiment, an atom (or other particle) is split, and its paired electrons (or photons) are sent in opposite directions. The spin of one electron (or the polarization of one photon) is measured, and then the other. It turns out that the second electron (or photon) always has the opposite spin (or polarization). The particles appear to be entangled. And their *entanglement* (a term coined by the famous physicist Erwin Schrödinger in his review of the original EPR experiment) occurs regardless of how far apart the paired particles are from each other when the measurement is taken. This is interesting but perhaps not too surprising. One might think of the paired particles as a foot-long stick that is broken into uneven pieces that are then separated. When one of the pieces is measured and found to be longer than six inches, the other piece will always turn out to be shorter than six inches.

The really strange thing happens when the spin (or polarization) of one of the distant particles is changed just before it is measured. Now the spin (or polarization) of the other particle, though miles away, immediately displays the opposite value. Somehow it immediately reflects the change in spin (or polarization) that its twin experiences, regardless of their distance from each other. It is as if the person holding the longer of the two sticks broke off a small piece just before measuring it, only to learn that the shorter stick miles away had magically grown by just this amount.

Nothing should be able to travel faster than light. Yet the measurements of entangled particles' spin or polarization seem to indicate instantaneous communication. Einstein aptly called nonlocality "spooky action at a distance." With EPR experiments, the curving of spacetime cannot come to the rescue, as it did to explain the force of gravity, for there are no massive bodies involved to curve space and time. In any case, the particles in question are far too distant for gravity, or any other such force, to have an effect. As physicist Henry Stapp states, "The EPR-type phenomena apparently entail the need for strong instantaneous influences, at some deep level, and this evidently entails, in turn, the need for a major restructuring of our ideas about the fundamental nature of the physical universe."[25] Nonlocality demonstrates a level of interdependence between elementary particles that defies understanding.

Physicists create complex mathematical formulas to inform their theories of nonlocality. And even as these competing theories get debated, quantum computers that exploit the physics of entanglement and nonlocality to gain speed and power hitherto unavailable are being developed.[26] The knowledge and skill required to engage in such highly sophisticated mathematics and engineering are considerable. Still, we might better understand the phenomenon of entanglement with a rather simple notion that physicists employ: the grid. A pervasive medium, called a quantum field or grid, is said to connect everything in the cosmos. As Nobel Laureate physicist Frank Wilczek observes, the grid is the "primary ingredient of physical reality" and forms the basic structure of spacetime.[27] The grid is everywhere, and nothing escapes it. The "empty space" that exists between planets, stars, and galaxies, like the "empty space" found between elementary particles, is not really empty. Rather, it is part of a grid bubbling with quantum activity. Particles, whether viewed as bits of mass or congealed energy, are understood to constitute "localized disturbances" of this omnipresent grid.[28]

For the most part, all we humans can perceive are the localized disturbances. We see only those places on the quantum grid where something called a singularity arises. But the basic idea is that the grid itself, not the localized disturbances, is the most fundamental reality. The visible universe, which appears to consist of interacting bits of matter, is simply the most prominent way that the grid manifests itself to us. Beneath all these perceivable and measurable singularities that give us the impression of separate entities and separate events, physicists suggest, the universe exists as "an undivided and unbroken whole."[29]

The grid spontaneously demonstrates particle transformations known as quantum fluctuations. Particles come into existence and disappear on a regular basis for no particular reason. Of course, this does not mean that rocks, trees, people, or entire planets simply appear one day and disappear the next. We are speaking of a quantum grid, so the spontaneous fluctuations occur primarily or wholly at the subatomic level. While particles may spontaneously manifest themselves as localized disturbances in the grid, on average and in the aggregate these appearances will be offset by a roughly equal number of disappearances. At the macroscopic level, life goes on as normal, abiding by the laws of classical mechanics. One might imagine quantum fluctuations to operate like the hydrology of an ocean. Each day countless molecules of water evaporate "spontaneously" from the surface of the seas. But an equal number, on average, fall back into the ocean in drops of rain. Constant flux at one level yields enduring stability at another.

In the Newtonian, mechanistic world, motion is continuous and matter is discrete. This is the world we typically perceive, where bodies of matter have stable, definable boundaries—whether they are continents, cannon-balls, or carbon-dioxide molecules—and demonstrate fluid movement. In this billiard ball world, the paths of moving objects are smooth, and the edges of the objects are hard. In the quantum world, the opposite is true. Motion is discontinuous, and matter is fungible. Here, bits of matter with no precise boundaries are readily transformed into packets of energy. Their motion or transformation is not smooth; it always occurs in distinct steps. The transition between stationary states or levels of energy occurs in discrete bundles or quanta.

This was Max Planck's great discovery at the onset of the twentieth century. Planck observed that the spectrum of light emitted from hot, glowing objects occurred in steplike increments. The spectrum was not a continuum but a series of distinct, separate colors. He derived a mathematical constant, now known as Planck's constant, to account for these minute steplike quanta. A few years later, in 1905, Einstein noted that only light of certain frequencies could create an electric current when shone on certain metals. For example, blue light produced a current but not red light. This effect occurred regardless of how dim the blue light or how bright the red light was. Einstein hypothesized that the electrons in the metal could receive energy from light only in discrete portions, or quanta, which he called photons. Only photons of a high enough frequency (blue is higher than red) were capable of knocking an electron free of its atom of metal to create an electric current. In 1921, Einstein received the Nobel Prize in physics for his theory of this photoelectric effect.

Einstein, like Planck, was a modern-day Pythagoras. Harmonic notes, the ancient Greek sage had observed, could be produced only in strings bearing lengths of particular ratios. Likewise, particles emit or absorb electromagnetic radiation, such as light, only in particular quanta. Energy, in other words, comes in distinct, finite packets. Less than a decade after Einstein's explanation of the photoelectric effect, Niels Bohr expanded the new science of quantum physics to include other atomic structures. Electrons circling a nucleus, Bohr demonstrated, have orbits of set radii. When an atom absorbs enough energy (e.g., a photon of a certain frequency), its electron will leap to a new orbital path with a greater radius. The electron does not make this transition in a slow, continuous movement as the energy is incrementally absorbed. Neither can the electron stop halfway en route to a new orbit. It is either in one orbit of a set radius or instantly in another. There is no in between.

At a quantum level, the universe is discontinuous, or granular. It is digital, one might say, rather than analog. When they were first developed, digitized music and photography could not achieve the same quality as, respectively, analog music and conventional film cameras. Once the digitization had reached a fine enough resolution, producing smoother and less grainy results, it became as good (to human ears and eyes) as its analog equivalents. The quanta or leaps between states in the subatomic world are small enough that they produce smooth continuities at the macroscopic level.[30] At the quantum level, however, Planck's constant makes a big difference. Indeed, if it were set to zero, the mathematics of quantum physics would become identical to the mathematics of classical Newtonian physics.

Since at least medieval times, if not earlier, it was understood that *natura non facit saltus*—nature makes no leaps. This rule adequately accounted for the continuities of the world that human beings saw, heard, tasted, and touched. The quantum world does not abide by this law. In the quantum world, electrons jump to higher orbits, and the spins of entangled particles affect each other without there being a continuous path of communication or movement. In a universe manifesting quantum fluctuations, one has to grapple with apparent nonlocalities (and noncausalities). It might seem that quantum theory, which maintains a discontinuous universe of particles engaging in "quantum leaps" would not be receptive to holism. After all, wholes are usually thought of as continuous rather than granular, as not making leaps. How, then, can a continuous whole be composed of discontinuous parts? We must return to the notion of a grid to understand.

The grid allows forms of connection other than those occurring when one object physically bumps in to another. In Newton's mechanistic world, change in motion (direction or speed) occurs only when the continuous path of one bit of matter intersects with the continuous path of another bit of matter, creating a direct impact that alters velocity. In this Newtonian world, action at a distance is impossible. But in a quantum grid, nonlocality can more easily be understood. Quantum leaps are fluctuations on the grid. Action at a distance is not an effect without a cause. Rather, it is the implicate order becoming explicate. A nonlocality, from this perspective, is like any other form of motion or change at the quantum level. In Bohm's words, it constitutes "a relationship of certain phases of *what is* to other phases of *what is*, that are in different stages of enfoldment."[31] Nonlocality is not the creation of something out of nothing, but rather the making explicit of something enfolded within the fabric of reality.

What exactly does it mean for the universe to be enfolded, to have an underlying implicate order? Bohm offers an illustration from fluid dynamics. Put some viscous liquid, such as treacle, in an open-ended cylinder, and insert a slightly smaller inner cylinder. Let a drop of insoluble black ink fall into the treacle-filled chamber between the concentric cylinders, and then slowly rotate the outer cylinder. The ink, pulled along by the moving treacle, will be stretched into a fine thread that eventually disappears as the cylinder continues to rotate. At that point, the chamber between the cylinders appears to be filled with a homogenous, slightly gray treacle. But this undifferentiated liquid actually bears the implicate order of the drop of ink. That is, the order (differentiation) of the droplet is enfolded within the seemingly homogenous treacle. If the outer cylinder is slowly turned in reverse, the thin thread of ink will reappear, grow thicker, and eventually reconstitute the original droplet.[32]

The implicate order can also be illustrated by a hologram. Holograms enable the viewing of images in three dimensions that bear the property of parallax. That is, holographic images change when viewed from different angles, just as the view of a three-dimensional object in space changes as one's angle of perception is shifted. With holograms, where you are standing largely determines what you see. A standard photographic image cannot mimic this feature of reality. Moving around to view a photograph from different angles does not change what you see.

Holograms have another unique property. If a photograph is cut into pieces, each piece will reveal a portion of the original image. When a piece of a holographic plate is broken off and illuminated, in contrast, it reveals an image of the original object in its entirety (though with reduced detail and from a particular vantage point). With holograms, each part contains the pattern for the whole. One might say that the form of the entire object is enfolded within every piece of the holographic plate. This form gets unfolded, and the implicate order revealed, when the holographic plate, or some portion of it, is illuminated by laser light.

Bohm suggests that the universe can be understood as a kind of hologram, where each point in spacetime is connected to and enfolds every other point. The implicate order of the universe goes beyond whatever explicate order is demonstrated when we observe bits of matter bumping into each other. The whole is contained within each of the parts. Holograms and the movement of treacle in cylinders do not so much demonstrate an implicate order as provide analogies of it. As Bohm makes clear, real enfoldment occurs in a "higher-dimensional space."[33]

David Bohm's notion of an enfolded universe, as an interpretation of nonlocality and quantum fluctuations, remains controversial. Empirical evidence is scant. To speak of the implicate order, scientifically speaking, is to skate on thin ice. Many more discoveries, and much more mathematics, need to occur before it can be confirmed. And it is quite possible that an enfolded universe at the subatomic level will never fully manifest itself in the macroscopic world. Nonlocality may be a quantum event with limited impact at the human scale. At the same time, the physics of an entangled cosmos is steadily growing, with concrete applications in development. It would be foolish to ignore such powerful manifestations of the "essential interconnectedness of the universe."[34]

Physics and Metaphysics

Physicist Richard Feynman once observed that "if you think you understand quantum theory, you don't understand quantum theory."[35] Physics
offers us a world of continuity (waves) and discontinuity (particles), of
certainties (Planck's constant, the speed of light in a vacuum, the gravitational constant) and uncertainties (the position, momentum, and even
existence of subatomic particles, and the nature of the most fundamental
cosmic forces). It is a world of tremendous power *and* paradox. To lose sight
of the paradoxical quality of the universe is to deny all the connections
we have yet to perceive and understand.

Most sciences, including physics, consist of investigations of specific
strands on Indra's net. Advances are made by uncovering connections
between two or more familiar elements previously thought to be unrelated
or between known and newly discovered elements. Throughout the history
of physics, scientists have hoped, and not infrequently announced, that
they were on the verge of discovering a unified "theory of everything" that
would account for all of the basic forces in the universe and unite its fundamental laws.

Some theoretical physicists believe that string theory will explain all.
They posit the existence of energetic "strings," or membranes, that are
billions of times smaller than the smallest particles now known. String
theory is meant to explain all the known forces of the universe and
describe their relations of interdependence. The hypothesized vibrating
strings or membranes are so small that they do not exist within the three-
dimensional reality we can see and touch. Mathematical formulas positing
a mind-boggling eleven dimensions are required to account for them.

Other physicists believe that the most expensive experimental effort in
the history of science will soon produce a theory of everything. Five
hundred feet below the Franco-Swiss border, near the city of Geneva,
protons and other high-energy particles are being accelerated in opposite
directions around a 17-mile circular tunnel. Sophisticated instruments
record the forces that are produced when these particles, traveling near the
speed of light, collide. Constructed and operated by the European Organization for Nuclear Research (CERN) at a cost of $9 billion, the Hadron
supercollider is the largest particle accelerator ever built. It is likely that
CERN's monumental feat, a collaborative effort of more than 10,000 sci-

entists from over 100 nations, will yield discoveries that require a restructuring of basic physics.

Physicists once thought the atom was an indivisible particle, only later to discover protons, neutrons, and electrons. Subsequently they discovered that protons and neutrons were made up of ever-smaller components, which they called leptons, baryons, mesons, quarks, gluons, gravitons, and bosons. Even if the Hadron supercollider discovers a more fundamental entity, such as the elusive Higgs particle, or string theorists account for ultrasmall membranes, we can be fairly confident that a still more fundamental energetic relationship remains unseen. Something always eludes our most heroic efforts to comprehend. William James's assertion comes to mind: "Everything you can think of, however vast or inclusive, has on the pluralistic view a genuinely 'external' environment of some sort or amount. Things are 'with' one another in many ways, but nothing includes everything, or dominates over everything. . . . However much may be collected, however much may report itself as present at any effective centre of consciousness or action, something else is self-governed and absent and unreduced to unity."[36] James suggested the term *multiverse* to describe a pluralistic cosmos that could never be reduced to a unity. At the same time, he insisted that "our 'multiverse' still makes a 'universe'; for every part, tho it may not be in actual or immediate connexion, is nevertheless in some possible or mediated connexion, with every other part however remote." Not all connections are "actualized at the moment," James argued. But they exist as potentialities, enfolded, as it were, in the universe.[37]

To be sure, the universe is a wild and wooly place filled with a multitude of connections, including many that remain mysterious. If the ecological, ethical, technological, economic, political, and psychological interdependencies that we navigate in our daily lives are so numerous and complex as to leave us baffled, consider a few of the relationships that physicists confront.

The most basic foundations of our mundane world, space and time, are not distinct, stable, and independent phenomena. Atomic clocks placed on high-altitude aircraft empirically demonstrate what theorists have mathematically proven: time slows down as matter accelerates in space. In turn, matter is not the enduring stuff of our everyday life, but rather a transitory form of congealed energy. Moreover, matter exists alongside antimatter. Antiparticles have been observed in cosmic rays and can be produced in laboratories. Particle-antiparticle pairs (particles of the same mass but opposite charge) can arise spontaneously in quantum fluctuations.[38] When they collide, they annihilate each other, producing photons, energy without mass. And some photons, along with other elementary particles, are entangled. They have mirror images with which they dance, ignoring the constraints of space and time.

Some of the best minds in physics today propose that the universe as a whole exists as a series of mirror images. In an effort to explain quantum fluctuations, theoretical physicists posit parallel universes, where each of the infinite number of possibilities suggested by probability waves effectively lives out its separate life. As we observed, the velocity and position of subatomic particles cannot be definitely determined. They occupy a wide range of positions and velocities simultaneously, a so-called quantum superposition described by a wave function. Theoretical physicists suggest that superposition describes not only the spacetime coordinates of the smallest particles, but the overall structure of the universe. If this is true, then the universe we detect and measure is only one of an infinite number of universes existing simultaneously. The theory resolves some of the uncertainties of quantum particles, but it does so by hypothesizing entirely new universes constantly coming into being to account for every possible position and velocity of every particle.

Although a virtually infinite number of parallel universes may exist, we could not perceive them from within the confines of our universe. Even within our own cosmos, we cannot perceive most of what exists. Just as visible light can easily pass through glass, so there are other subatomic particles that pass through other types of matter with little or no resistance. These particles constantly bombard us. About 50 trillion neutrinos pass through our bodies every second, and we are none the worse for it. Intergalactic space was once assumed to be empty. We now know that it is filled with electromagnetic radiation. In turn, intergalactic space is rife with what physicists call dark matter and dark energy, both of them invisible to our best telescopes. We know of their existence only by way of the force they exert on nearby stars and galaxies.

Dark energy and dark matter are not negligible forces. According to current estimates, normal matter, the stuff we can see and measure, constitutes about 5 percent of the universe, while the rest is made up of dark energy (70 percent) and dark matter (25 percent).[39] Dark matter forms diffuse halos around galaxies that may weigh five times as much as the galaxies themselves. It provides a gravitational force pulling the universe together. Dark energy, in contrast, pushes the universe apart, or, rather, it pushes itself apart. Dark energy is not so much *in* space as it is creative *of* space. It appears to be uniformly distributed throughout the universe. What we know of dark energy and dark matter confirms that most of the cosmos is concealed from us. As Heraclitus said, nature loves to hide.

It is not unreasonable, in light of such enigmas, to assume that vast hidden forces structure our world. The psychologist Viktor Frankl once criticized reductionism as a form of "nothingbutness." It is fine and good to maintain that the whole can be divided into ever smaller parts and that the effects of these parts bumping into each other explain a great deal. But we should not insist that any particular phenomenon—be it the movement of a subatomic particle or the action of a human being—is *nothing but* the mechanistic interaction of smaller parts. That would mistake the current limits of our minds and measurements for the boundaries of reality. Whether an implicate order pervades our wild and woolly universe (and all the parallel universes that may exist) remains unknown. But one thing is clear: whether carried out on the shores of ancient Greece, from telescopes hurtling through outer space, or in minute particles flying near the speed of light underneath Geneva, metaphysics and physics have taught us the same lesson: the cosmos is much more than meets the eye.

Conclusion

In 1514, Nicolaus Copernicus produced a six-page, handwritten document that claimed the planets, including the earth, were circling the sun. Over the next three decades, the Polish astronomer refined his heliocentric hypothesis, eventually publishing *On the Revolutions of the Celestial Spheres* just before his death. Copernicus's groundbreaking work was well received by both lay scholars and the church authorities, including the archbishop of Capua. Galileo Galilei would not be so fortunate. Seven decades after Copernicus's death, in 1616, Galileo was given the ultimatum by Pope Paul V to renounce his belief in heliocentrism or be branded a heretic and face the consequences. Galileo agreed to abandon his belief. But he was back on trial for heresy in 1633, again for promoting the Copernican hypothesis. Again he was forced to recant, after which his publications were banned. The Italian astronomer passed the remainder of his days under house arrest.

The fates of Copernicus and Galileo are remarkably different, notwithstanding their similar assertions. Church authorities were tolerant of Copernicus's claim, even quite admiring of it as a feat of intellectual prowess. Galileo was violently castigated for speaking contrary to Holy Scripture; he was forced to recant and prevented from carrying out his studies. Why the difference? Galileo had transformed theory into practice. Whereas Copernicus worked at the level of hypothesis and mathematical computation, Galileo gathered empirical evidence. The Italian astronomer confirmed the Copernican thesis by way of telescopic observations that could be repeated and verified. He saw the moons of Jupiter. And if these satellites orbited around a distant planet, then there were some things that did not revolve around the earth. While astronomical theory was tolerable for the church, the fact of the earth's not being the center of the universe was not. Galileo's empirical demonstration was too direct a threat to the authority of scripture and its papal interpreters. The Copernican revolution, as a worldly event, did not start with its namesake but with Galileo, its first victim.

One might say the same thing about the physics and metaphysics of interdependence and holism. For thousands of years, as the tale of Indra's net and the speculations of pre-Socratic philosophers attest, the universe was conceived as a web of relationships with the whole reflected in each of the parts. Such theories of interdependence and holism were gestures toward the unseen. As such, their implications were easy enough to ignore, and beyond stimulating metaphysical conjecture, their impact was quite limited. A competing reductionist theory of a deterministic world of distinct objects bumping and crashing into each other, at least since the days of Newton, has enjoyed something close to scriptural authority. Its success is indisputable. Technology and engineering based on reductionism has transformed our world.

In the past two centuries, however, the mechanistic worldview has been soundly rebuffed by empirical observations. Newton's corpuscular theory of light was destroyed by the double-slit experiment that Young performed in the early 1800s and the subsequent discoveries of wave-particle duality. Newton's claim that space, mass, force, and time are the fundamental, independent, stable building blocks of the cosmos was countermanded by Einstein's theories of special and general relativity in the early 1900s. Later in the century, atomic bombs, nuclear reactors, and atomic clocks empirically validated Einstein's assertions. Newton's deterministic and predictable world, best described by Laplace, was undermined by chaos theory and the vivid demonstrations of emergence over the past half-century. Finally, the Newtonian world of purely local actions has been shattered by experiments in the past few decades that demonstrate the entanglement of particles and nonlocality.

Our very effort to understand and measure the world, quantum physics demonstrates, implicates us in its dynamic order. Reflecting on this phenomenon, John Wheeler, the famous Princeton physicist, observed that "useful as it is under everyday circumstances to say that the world exists 'out there' independent of us, that view can no longer be upheld. There is a strange sense in which this is a 'participatory universe.'"[40] Interdependence and interpenetration are the most salient features of a participatory universe.

It is possible that the Hadron supercollider will have as revolutionary an effect as Galileo's (much less expensive) invention, the telescope. Looking through his optical device, Galileo demonstrated that the earth is not the stable, central, and unmoving foundation perceived by the naked eye. The supercollider may demonstrate that the universe is an interconnected grid. Of course, it may turn out to be the biggest bust in the history of physical experimentation. Yet we might hope that it provides further evidence of a cosmic web that connects everything to everything else. To be sure, we must be wary of the capacity of metaphysical hopes to conceal our fundamental ignorance. So it is crucial that we empirically test our theories and hypotheses in rigorous ways. Specifically, we must explore the many different ways in which things are connected, the relative significance of these connections, and the consequences of severing them.

Of course, readers should be wary of nonspecialists (like myself) who might cherry-pick concepts, terminology, and experimental evidence from physics or any other discipline to serve their own purposes. Admittedly, there is a thin line between ill-informed opportunism and interdisciplinary scholarship. A popular author insists that "each of us is a different person in different places. This doesn't make us inauthentic; it merely makes us quantum."[41] If the point is simply that context matters, well, that is harmless enough. Quantum physics teaches us that context matters, as do biological science, social science, humanistic fields of study, and worldly experience. Each of us can, and often should, act differently in different situations. At the same time, it would be intellectually dishonest to shield the vices of hypocrisy and fickleness behind the halo of quantum science. There may be many good reasons for abandoning a practical plan or a moral principle in a specific context; the superposition of subatomic particles is not one of them.

It is all too easy for us to give credence to a claim about the world simply because it buttresses our personal values or preferences. To be fair, however, nonspecialists with a penchant for holism are not the only people susceptible to such temptations. In a plea for objective science within the context of a critique of Bohm's implicate order, physicist Victor Stenger writes:

I ask myself: Do I really want to be one with the universe, so intimately intertwined with all of existence that my individual existence is meaningless? I find I much prefer the notion that I am a temporary bit of organized matter. At least I am my own bit of matter. Every thought and action that results from the remarkable interactions of my personal bag of atoms belongs to me alone. And so these thoughts and actions carry far greater value than if they belonged to some cosmic mind that I cannot even dimly perceive.[42]

Stenger suggests that his dismissal of an implicate order is grounded on the loss of a sense of self-worth that such a scientific finding might provoke. That I might find the idea of being "intimately intertwined with all of existence" consoling while Stenger finds it abhorrent testifies to human plurality. Neither orientation should impede rigorous scientific investigation, wherever it leads.

Carl Jung, like many other intellectuals of his day, was intrigued by the discoveries of science. He met with Einstein on numerous occasions, and corresponded with Wolfgang Pauli, one of the founders of quantum physics, for more than two decades. Observing the need for an "energetic" rather than Newtonian standpoint, Jung explored the "complementarity" of psychology and physics, noting that Heisenberg's work reinforced the psychological principle of the impact of observer over observed.[43] Jung's followers have continued to use physical science as a means of illuminating psychological reality. Just as light manifests itself as a wave-particle duality, Jungian scholar Edward Edinger writes, so people should be understood as *"both* unique individual units of being and also part of the continuum which is the universal wave of life."[44] This is a provocative metaphor.

Wrapped up as we are in our professional lives, no less than in our personal challenges and aspirations, we tend to see ourselves as particles, distinct from the medium through which we pass, separate from the environment in which we live. We perceive ourselves as individuals and gain a sense of power and purpose from this perspective. Still, our individuality, like the particle nature of a photon, is only half the story. While light acts like an individual particle, it also displays itself as a wave, an energetic relationship within the cosmological grid. We must always distinguish between inspiring analogies and an actual physical correspondence. At the same time, it would be strange indeed if the deepest levels of physical reality did not teach us something profound about who and what we are.

In his metaphysical writings, Arthur Koestler observed that the mind-boggling nature of quantum physics should make us more open to invisible connections within our world.[45] Koestler coined the term *holon* by combining the Greek word *holos*, which means whole, with the suffix *on*, which was meant to suggest a particle or part (as in the parts of an atom: the neutr*on*, prot*on*, and electr*on*).[46] Complex systems, Koestler proposed, are composed of holons, separate entities defined by the whole of which they partake. The cosmos is a *holarchy*, a set of nested holons. In our holarchic universe, Koestler maintained, all parts are connected to, and incorporate, the whole. Everything exists in relation.[47] On Indra's net, in other words, it's turtles all the way down.

Albert Einstein observed that as individuals, we necessarily experience ourselves as "limited in time and space" and as "something separate from the rest." But an individual's sense of separateness is "a kind of optical delusion of his consciousness. . . . Not to nourish the delusion but to try to overcome it is the way to reach the attainable measure of peace of mind."[48] Perhaps Einstein was speaking here more as a metaphysician than as a physicist. Perhaps the two fields are not so separate.

In any case, freeing ourselves from the delusion of separateness is the mandate of ecosophic awareness. And ecosophic awareness should be an impetus to deepen and broaden science, not an excuse to curtail or abandon it. In addition to the beliefs we employ to buttress a personal sense of self-worth or to attain peace of mind, we ought to embrace as well empirical studies and experimentation. Open-minded inquiry and rigorous science are crucial to the cultural creativity and adaptive change that will allow our species to survive and prosper. Exploring our selves and our world as connected strands of the cosmic web is the grandest adventure, and a homecoming.

8 Conclusion

We shall not cease from exploration
And the end of all our exploring
Will be to arrive where we started
And know the place for the first time.
Through the unknown, unremembered gate
When the last of earth left to discover
Is that which was the beginning.
—T. S. Eliot

With hands that grasp, minds that plan, and hearts that crave, human beings have made claim to dominion of an entire planet while thrusting themselves into the galaxy. We might rightly wonder if there are any limits to our species' ambitions and power. The ancient Greek playwright Sophocles had an answer to this question. The clever and ambitious Oedipus receives this counsel: "Do not seek to be master in every way." The pursuit of sovereign power brings doom. Sophocles acknowledged that human beings are "clever beyond all dreams" and the most "wondrous" of creatures. Our "inventive craft," however, wears away the ageless earth and drives us "one time or another to well or ill." In the face of such circumstances, Sophocles concludes, enduring happiness depends on wisdom.[1] Without wisdom, the knowledge and acceptance of limitation, tragedy is inevitable. Sustainability, like happiness, depends on wisdom. The alternative to its dedicated pursuit is tragedy on a global scale.

The pursuit of sustainability is challenging for the same reasons that make it necessary. Our world is holarchic, defined by nested complex systems. Its navigation is troubled by the law of unintended consequences. Net dwellers like us are both constrained and empowered by the relations of interdependence and interpenetration that surround us. Even the least of our actions sends ripples across the gossamer threads of Indra's net. Ubiquitous connections ensure significance for everything we do and everything we fail to do. This awesome fate can prove debilitating or empowering. It is debilitating for those who crave mastery in every way, empowering for citizens invested in community.

This book will not have merited the resources it incorporates if its readers are not stimulated to think and act differently. The life of the mind is a glorious thing, and I consider myself a staunch advocate of it and the ivory tower that provides it a secure *oikos*. At the same time, I agree with John Dewey that it is counterproductive "to emphasize states of consciousness, an inner private life, at the expense of acts which have public meaning and which incorporate and exact social relationships."[2] Dewey counsels us to "fix our attention" on developing the social conditions that facilitate the cultivation of the right sort of "habits," namely, those habits that promote creative thought, adaptive behavior, and responsibilty. This is the task of education and legislation, of cultural development, social policy, and community building. It is a task for mind and body or, rather, for embodied minds and mindful bodies.

I have employed the term *ecosophic awareness* to capture the attention-fixing, habit-forming, mindfully embodied sensibility of net dwellers. Ecosophic awareness prompts a responsive involvement in holarchy, from the inner polity with all its shadows to our social and ecological habitats. Ecosophic awareness is not only a humbling acknowledgment of limitation. It is also a stimulant of creativity, a provocation of freedom, and a catalyst for community. It strips from us the illusion that we could ever master life and control the future. But it does not, for that reason, discount efforts to improve our lot. Rather, it constitutes an attunement to the responsibilities and opportunities that a connected world presents.

To say that everything is connected is to utter something of a truism, a fact of life reinforced by the most casual glance at the world around us. For all the talk of interdependence in an age of globalization, however, the term is often misunderstood. Interdependence is about relationship, to be sure, but it pertains to a certain kind of relationship. It is neither independence nor dependence.

When relationships of dependence gain the upper hand, a rigid, non-resilient system develops. Such a system may function with great efficiency when everything is up and running, like a mechanical clock with all parts intact. But when something goes wrong, when a key element malfunctions, breakdown is inevitable and usually catastrophic. Of course, given an ever-changing world, something always goes wrong. A system grounded in dependence, like a chain, is only ever as strong as its weakest link. The failure of a single component causes it to fly apart.

When relationships of independence gain the upper hand, the threat is different. No man, or any other organism or institution, is an island, entire of itself. Individuals who aim for island-like autonomy, like the proverbial hermit, at best may derive from the politics of the soul some of the benefits of relationship that they have forgone in the world. The effort to manifest a full-blown state of autonomy, the Greek tragedians knew, was best left to the gods. In social settings, in any case, one person's rise to despotic autonomy marks another person's fall into disempowering servitude. Like the recluse, both the despot and the dominated (though seldom to equal degree) suffer from a lack of communication, cooperation, and productive competition.

Interdependence marks a fluid balance between independence and dependence. Often an increased appreciation of our connectivity is most needed. At other times, the development of realms of (relative) autonomy is called for, lest we become too dependent on rigid, monopolistic systems lacking in resilience. There is no hard-and-fast rule to determine how the balance of interdependence can and should be struck. Practical wisdom informed by the best science is all we have.

Six years after meeting with utopian socialist Henri de Saint-Simon, Barthélemy Prosper Enfantin (1796–1864) established a short-lived commune in Ménilmontant, on the outskirts of Paris. Here Enfantin and his forty disciples worked together while singing hymns to social harmony. They also wore clothes with all buttons sewn on the back. The idea behind this sartorial requirement was that members of the commune could not even get dressed in the morning or disrobe at night without relying on the good graces of their comrades.[3] Undoubtedly this exercise increased the utopians' appreciation of mutual need. But relationships of interdependence already exist in abundance and need not be artificially created for pedagogical purposes. In any case, we can and should experience, and enjoy, our limited independence as individuals, families, communities, and nations. Cultivating our autonomy *and* our responsible attunement to the relationships that embed us within larger webs of life is the meaning of ecosophic awareness.

Ecosophic awareness finds its concrete application in the practice of solving for pattern. Given our general state of ignorance regarding the relations of interdependence that pervade our ecological, ethical, techno-logical, economic, political, psychological, physical, and metaphysical lives, being on the right track will often be the product of fortune. But solving for pattern can and should become a conscious exercise.

If Americans would walk or cycle for thirty minutes a day rather than using their cars, carbon emissions would be cut by some 64 million tons a year and significant natural resources would be conserved. In addition, over 3 billion pounds of fat would be trimmed from bodies increasingly susceptible to obesity, diabetes, stroke, and heart attack.[4] And that is to say nothing of the reduction in noise, the multiplication of opportunities for social connection, the decrease in pollution, and the conservation of land and beautification of neighborhoods that would result if roads and parking lots gave way to walkways and bike paths. The more we know about the relationships of interdependence within our world, the better we get at solving for pattern.

For an activity, enterprise, or project to be sustainable, it must be eco-logically safe, economically viable, and socially equitable, while simultane-ously fostering creativity and learning. By integrating the pursuit of these diverse but interdependent goods, sustainability prompts us to solve for pattern. In the early 1960s, President John F. Kennedy observed, "Anyone who can solve the problems of water will be worthy of two Nobel prizes—one for peace and one for science." Kennedy was aware of the likelihood of conflict in hot spots around the world amid dwindling resources of freshwater. What he said of water half a century ago is true today of a wide range of natural resources whose rapid depletion will beget political turmoil and threaten military conflict. Our problems are highly interdependent, and their solutions should be no less so. Indeed, anyone who achieves a major breakthrough in the arena of sustainability today deserves numerous Nobel prizes: for peace, which is grounded on social justice and equity, for economics, and for creativity and learning across scientific and cultural disciplines.

Sustainability in Action

To pursue sustainability and solve for pattern is not to fashion or implement a static blueprint. Our survival, not to mention our hope for prosperity, depends on adaptation. Sustainability requires that we solve for pattern in an ongoing, creative and collective enterprise of adaptive management and synergistic coevolution. The conviction that all the parts are connected is crucial. But equally imperative is the effort better to discern— employing the best science and the full scope of our embodied intelligence—how all the parts are connected. This entails linking diverse stakeholders into networks that gather and assess information, engage in safe-to-fail experimentation, monitor outcomes within and across various scales, and learn through practice.[5]

Experimentation may seem incongruent with sustainability. But in a world whose only certainty is change, adapting—at the proper scale and speed—is the only means to sustain what we value. Today we are changing our planetary *oikos* far more rapidly than we are gaining understanding of it. Rapid change and too-big-to-fail institutions are an invitation to catastrophe. As Schumacher observes, "There is wisdom in smallness if only on account of the smallness and patchiness of human knowledge, which relies on experiment far more than on understanding. The greatest danger invariably arises from the ruthless application, on a vast scale, of partial knowledge."[6] All too often, we exercise cleverness on a large scale and accelerated pace to compensate for the wisdom and patience that elude us. A technological breakthrough a day, we hope, will keep our ecological and social crises at bay. In contrast, adaptive management entails learning how to manage while managing to learn. Its ecologically attuned efforts are studies in self-limitation. For Aristotle, like Dewey, virtue was something one could learn only by doing. Likewise, we come to know the practical meaning of sustainability by living sustainably, ever learning from the repercussions that issue from our tentative efforts.

We live in an always-changing world defined by fluid relationships. Everything is in flux—from the realm of subatomic particles and the protean world of human interactions to galactic streams of curved space and time. From quanta to community to cosmos, we live in webs of connected yet ever-shifting relations. Sustainability is not a final state of rest or changelessness. Rather, its pursuit helps us manage change. It is an evolving path, not a destination.

Aldo Leopold wrote, "We shall never achieve harmony with land, any more than we shall achieve absolute justice or liberty for people. In these higher aspirations the important thing is not to achieve, but to strive. It is only in mechanical enterprises that we can expect that early or complete fruition of effort which we call 'success.'"[7] In a technological society captivated by mechanical enterprises and Promethean values, the prospect of an endless venture in sustainability might seem discouraging. It should be inviting. Lao Tzu insisted that "a good traveler has no fixed plans and is not intent on arriving." Sustainability is a journey that demands, and helps to cultivate, good travelers.

In becoming better travelers, we might take our lead from nature itself. After all, evolution is an adventurous journey forever uncovering fresh landscapes. The eighteenth-century German polymath Johann Wolfgang von Goethe observed, "If we want to achieve a living understanding of nature we must follow her example and become as mobile and flexible as nature herself."[8] How could it be otherwise? We have neither the knowledge nor the power to prevent change. Nothing in this world—neither landscape nor continent, neither organism nor ecosystem, neither individual nor society—is permanent and fixed.

Heraclitus's truth, ironically, has withstood the test of time. We cannot step into the same river twice. But the transient, chaotic, and unpredictable nature of a river need not keep us from building just communities on its banks, drinking its life-giving water, or caring for its web of life. Indeed, our appreciation of its fundamental flux—appreciation grounded in ecosophic awareness—facilitates living sustainably on its shores.

Assessing the transitioning ecologies of my own adopted state, C. S. Holling and Lance Gunderson observe:

Some ten thousand years ago (very recent in geologic time frames) the treasured Everglades of southern Florida were not wetlands, but a dry savanna. Had we been living then, would we, as people concerned with the conservation of nature, have sought to maintain that savanna state as desirably pristine, holding back the rising seas as glaciers melted? Placing fingers in the dikes we built? Denying the reality of climate change? Is it desirable to have a goal of preserving and protecting systems in a pristine, static state? . . . Efforts to freeze or restore to a static, pristine state, or to establish a fixed condition are inadequate, irrespective of whether the motive is to conserve nature, to exploit a resource for economic gain, to sustain recreation, or too facilitate development. Short-term successes of narrow efforts to preserve and hold constant can establish a chain of ever more costly surprises.[9]

Holding constant and preventing change, even on those rare occasions where it may be possible, is never doing just one thing. It will produce unintended consequences. With this in mind, Holling and Gunderson observe, "The challenge, rather, is to conserve the ability to adapt to change, to be able to respond in a flexible way to uncertainty and surprises. And even to create the kind of surprises that open opportunity. It is this capacity that a view of an evolving nature should be all about—i.e., maintaining options in order to buffer disturbance and to create novelty."[10] To pursue sustainability is to foster resilience.

Michel de Montaigne observed that "athwart all our plans, counsels, and precautions, fortune still maintains her grasp on the results."[11] We have not tipped the balance in favor of our control of the future in the 400 years since Montaigne gave voice to his renowned skepticism. But a healthy doubt of humanity's power to master fortune does not absolve us of the responsibility to act in ways that preserve our better values and portend a better future. For the most part, we do not know what we are doing or what we ought to do. But we cannot rest in passivity, transfixed by doubt, because inaction also has consequences, not the least for the nonactor. There will always be reasons to be found for inaction. But the ill effects of passivity are more numerous.

The author Vladimir Nabokov once wrote that "what can be controlled is never completely real; what is real can never be completely controlled."[12] We exist within a web of life. That is our reality, and it is complex. Complexity forbids control. Pursuing sustainability entails weaning ourselves from the illusion of control without, at the same time, starving ourselves of the nourishment of hope. Hope refers to the capacity, and predilection, to see the world as pregnant with possibilities.

The hope we need to cultivate today is not hope for a panacea. Quite the opposite. Anything big enough to serve as a cure-all is also big enough, and more likely, to produce unintended consequences of catastrophic dimensions. Were some inexhaustible, reliable, and vast carbon-free source of energy to be developed, we might rightly consider ourselves fortunate (or clever enough) to have dodged the bullet of global warming. Some suggest that hydrogen fusion—the energy that powers the sun, harnessed here on earth—will be our salvation. Even if fusion does help us slow down, stop, or even reverse climate change, we can be certain that it will open up a Pandora's box of other problems. But that is no reason to hinder its development. At this point, the climate crisis calls for all hands on deck. Yet the very prospect of a radical solution, like the problem of climate change itself, should prompt us to invest our ingenuity in more than technology. We need to dedicate equal amounts of effort to the building of resilient, ecosophic communities—at local, national, and international scales—so that might well assess, engage, and respond to the problem of climate change, its proposed "solutions" and their unintended consequences.

Most of the social and ecological problems we face are the side effects of actions that were proffered as solutions to previous problems. Our challenges today are massive and pressing. But practical wisdom suggests that there are no single solutions commensurate in size to the magnitude of our problems. As Wendell Berry observes, today's "great problems call for many small solutions."[13] Prudence demands that our actions be tempered by knowledge and that our knowledge be tempered by wisdom. Wisdom is knowledge aware of its own limitations. Wisdom, as Socrates demonstrated, is an appreciation of one's own ignorance.

Ignorance does not have to stymie the pursuit of sustainability. If our "solutions" are numerous and small enough, they and their unpredictable side effects can be part of an adaptive, experimental approach. "Since we're billions of times more ignorant than knowledgeable," Bill Vitek and Wes Jackson muse, "why not go with our long suit and have an ignorance-based worldview?"[14] Such a worldview presents sustainability as the challenge of living well with uncertainty. Speaking of the future of a connected world, Paul Hawken, Amory Lovins, and Hunter Lovins observe that "the most unlikely environmental scenario is that nothing unlikely happens. The biggest surprise would be no surprises."[15] Our current state of ignorance provides the justification—indeed the imperative—for creative, small-scale, safe-to-fail experimentation in the pursuit of more sustainable lifestyles, businesses, and societies.

Promising golden lives of justice, order, comfort, plenty, and power, our Midas touch delivers us into a realm of endless and escalating side effects. In this realm, yesterday's solutions become today's problems. And today's well-intentioned or self-serving solutions—consider current plans to combat disease with nanotechnology, modify domesticated species genetically to increase productivity, or geoengineer the atmosphere to offset global warming—will most certainly supply some of tomorrow's most pressing problems. In the business world, a person who has "the Midas touch" is understood to be either so skillful or so fortunate that any enterprise he undertakes becomes successful. That is a strangely myopic interpretation of the ancient myth. It bears in mind only half of the story of the wide-eyed king—the part before unintended consequences come home to roost. We so want to increase our power, wealth, and comfort in this world that we have transformed a myth designed to teach a sobering lesson into a celebration of shortsighted ambition. Effectively, the Midas touch has been applied to the tale of Midas.

Our incapacity to do just one thing may be perceived as a tragedy, just as our state of interdependence may be viewed as a curse. But that is a choice we make. Becoming better travelers on Indra's net means experiencing interdependence and the law of unintended consequences as opportunities. With this in mind, the challenge of sustainability refers not only to our relationship to a world in peril. It pertains equally to our inner lives and personal struggles. Plotinus, a third-century neo-Platonist philosopher, wrote, "Never did eye see the sun unless it had first become sunlike, and never can the soul have vision of the First Beauty unless itself be beautiful."[16] Only the person who perceives and celebrates the diverse, dynamic elements of his or her self can well navigate a pluralistic and protean universe. Once we realize that our every action leaves its legacy in us as well as in our world, our inability to do just one thing becomes empowering, a blessing rather than a burden. Thriving in a connected world demands the best of us.

Notes

Introduction

1. See the *KOF Index of Globalization*, http://globalization.kof.ethz.ch/, and the aggregate data from 1970 at http://globalization.kof.ethz.ch/query/. See also the A. T. Kearney/Foreign Policy Globalization Index, http://www.atkearney.com/index.php/Publications/globalization-index.html.

2. Arne Naess (1912–2009), the Norwegian philosopher who developed the concept of deep ecology, coined the term *ecosophy* in the early 1970s. For Naess, ecosophy designated a personalized philosophy of social and ecological harmony or equilibrium. Naess relied in part on insights from Mahayana Buddhism, particularly the notion of the self-realization of all beings, to develop his concept of "'finding oneself' . . . in deep connection to all that surrounds." Arne Naess, *Ecology, Community and Lifestyle: Outline of an Ecosophy* (Cambridge: Cambridge University Press, 1989), 80. The French theorist and psychoanalyst Felix Guattari also employed and refined the notion of ecosophy. For Guattari, ecosophy designated an ecological, ethical, and aesthetic theory and practice oriented to the decentered, heterogeneous, and nonreductionistic pursuit of social, mental, and environmental well-being. Felix Guattari, *The Three Ecologies*, trans. Ian Pindar and Paul Sutton (London: Athlone Press, 2000). Naess, and to a lesser extent Guattari, have contributed to my understanding of ecologically grounded worldviews. But I do not rely on or reference them in what follows, and my use of *ecosophic* does not correspond to their related term.

3. *Oxford English Dictionary* (New York: Oxford University Press, 1971), 891.

4. Theodore Roszak, "Where Psyche Meets Gaia," in *Ecopsychology*, ed. Theodore Roszak, Mary Gomes, and Allen Kanner (San Francisco: Sierra Club Books, 1995), 8.

5. Paul Hawken, Amory Lovins, and L. Hunter Lovins, *Natural Capitalism* (New York: Little, Brown, 1999), 8, 14. Also see the 2008 statistics from the Environmental Protection Agency site on municipal waste, http://www.epa.gov/osw/nonhaz/municipal/msw99.htm

6. See "Carbon Footprint of Best Conserving Americans Is Still Double Global Average," *Science Daily*, April 29, 2008, http://www.sciencedaily.com/releases/2008/04/080428120658.htm.

7. Francis H. Cook, *Hua-yen Buddhism: The Jewel Net of Indra* (University Park: Pennsylvania State University Press, 1977), 2.

8. Ibid., 14, 42.

9. The Indian Avatamsaka Sutra is the central scripture of the Hua-yen tradition of Buddhism. Indra's net is specifically mentioned there. See *The Flower Ornament Scripture: A Translation of the Avatamsaka Sutra*, trans. Thomas Cleary (Boston: Shambhala, 1984), 226, 389. Tu-shun tells the story of Indra's net in a more organized fashion. See Tu Shun, "Cessation and Contemplation in the Five Teachings of the Hua-yen," in *Entry into the Inconceivable: An Introduction to Hua-yen Buddhism*, trans. Thomas Cleary (Honolulu: University of Hawaii Press, 1983); see also Cook, *Hua-yen Buddhism*, 2.

10. Cook, *Hua-yen Buddhism*, 2.

Chapter 1

1. Garrett Hardin, "The Cybernetics of Competition: A Biologist's View of Society," *Perspectives in Biology and Medicine* 7 (1963): 58–84, reprinted in Garrett Hardin, *Stalking the Wild Taboo* (San Francisco: Morgan Kaufmann, 1978).

2. Garrett Hardin, *Living within Limits* (New York: Oxford University Press, 1993), 199.

3. Robert K. Merton, "The Unanticipated Consequences of Purposive Social Action," *American Sociological Review* 1, no. 6 (1936): 894–904.

4. While taken to be synonyms in this book and in general parlance, the terms *unintended consequences* and *unanticipated consequences* have distinct meanings. The former denotes *unwilled* effects and the latter *unexpected* effects. Unintended consequences may be anticipated, and unanticipated consequences may be intended. Military forces, for example, may not intend to harm civilians when bombing enemy forces. Civilian casualties are unintended in the sense that noncombatants are not targeted. At the same time, military forces may fully anticipate that there will be a number of civilian deaths, euphemistically called "collateral damage," when they drop bombs in populated areas. In the same way, unanticipated consequences may be intended. For example, when a pessimist (or realist) buys a lottery ticket, she might rightly say that winning the jackpot is unanticipated. She fully expects *not* to have the winning ticket, and would predict as much. But her intention in buying the ticket is to win.

5. Merton, "The Unanticipated Consequences of Purposive Social Action," 900.

6. Rob Norton "Unintended Consequences," 1, http://www.econlib.org/LIBRARY/Enc/UnintendedConsequences.html.

7. Lester Milbrath, "Environmental Understanding: A New Concern for Political Socialization," in *Political Socialization: Citizenship, Education and Democracy*, ed. Orit Ichilov, 292 (New York: Teachers College Press, 1990).

8. See International Union for the Conservation of Nature's Red List, http://www.iucn.org/about/work/programmes/species/red_list/about_the_red_list/.

9. *Discover Magazine*, December 2006, lists *Silent Spring* at number 16 of the top 25; http://discovermagazine.com/2006/dec/25-greatest-science-books/article_view?b_start:int=1&-C=.

10. Garrett Hardin, "An Ecolate View of the Human Predicament," in Hardin, *Stalking the Wild Taboo*.

11. Rachel Carson, *Silent Spring* (Boston: Houghton Mifflin, 1962), 278.

12. Mark Hamilton Lytle, *The Gentle Subversive: Rachel Carson*, Silent Spring, *and the Rise of the Environmental Movement* (New York: Oxford University Press, 2007), 165.

13. Ibid., 175.

14. Quoted in Jonathan Norton Leonard, "Rachel Carson Dies of Cancer; 'Silent Spring' Author was 56," April 15, 1964; http://www.nytimes.com/books/97/10/05/reviews/carson-obit.html.

15. Bridget Stutchbury, *Silence of the Songbirds* (New York: Walker, 2007).

16. Donald Worster, *Nature's Economy* (Cambridge: Cambridge University Press, 1994), 284.

17. Gifford Pinchot, *The Fight for Conservation* (New York: Doubleday, 1910), 79.

18. Aldo Leopold, *A Sand County Almanac, with Essays on Conservation from Round River* (New York: Ballantine Books, 1966), 139–140.

19. Ibid., 240.

20. Ibid., 138–139.

21. Ibid., 190.

22. Morris Berman, *The Reenchantment of the World* (Ithaca, NY: Cornell University Press, 1981), 257.

23. J. A. Estes, D. O. Duggins, and G. B. Rathburn, "The Ecology of Extinctions in Kelp Forest Communities," *Conservation Biology* 3 (1989): 251–264.

24. See R. Lewin, "In Ecology, Change Brings Stability," *Science* 234 (1986): 1071–1073; Stuart Pimm, *The Balance of Nature?* (Chicago: University of Chicago Press, 1991).

25. Thomas Fritts and Gordon Rodda, "The Role of Introduced Species in the Degradation of Island Ecosystems: A Case History of Guam," *Annual Review of Ecology and Systematics* 29 (1998): 113.

26. Other sources indicate that the snake has eradicated twenty-three of Guam's thirty-four native bird species. See Gordon H. Rodda and Julie A. Savidge, "Biology and Impacts of Pacific Island Invasive Species," *Pacific Science* 61 (2007): 307–324.

27. Jason Van Dreische and Roy Van Dreische. *Nature Out of Place: Biological Invasions in a Global Age* (Washington, DC: Island Press, 2000).

28. Cornelia Dean, "Research Ties Human Acts to Harmful Rates of Species Evolution," *New York Times*, January 13, 2009, D3.

29. See F. Y. Cheng, "Deterioration of Thatch Roofs by Moth Larvae after House Spraying in the Course of a Malaria Eradication Program in North Borneo," *WHO Bulletin* 28 (1963): 136–137; G. R. Conway, "Ecological Aspects of Pest Control in Malaysia," in *The Careless Technology,* ed. M. T. Farvar and J. Milton, 467–488 (New York: Natural History Press, 1969); T. Harrisson, "Operation Cat Drop," *Animals* 5 (1965): 512–513.

30. Norman Myers, *Ultimate Security: The Environmental Basis of Political Stability* (New York: Norton, 1993), 204–205.

31. Lester Brown, *Plan B 3.0: Mobilizing to Save Civilization* (New York: Norton, 2008), 4.

32. Ibid., 56.

33. M. Scheffer, M. Holgrem, V. Brovkin, and M. Claussen, "Synergy between Small- and Large-Scale Feedbacks of Vegetation on the Water Cycle," *Global Change Biology* 11 (2005): 1003. See also Bjorn-Ola Linnér, "Authority through Synergism: The Roles of Climate Change Linkages," *European Environment* 16 (2006): 279, 286. See also American Geophysical Union Centre for Ecology and Hydrology, Dorset Wageningen University, Joint Release, May 22, 2006, http://www.agu.org/news/press/pr_archives/2006/prrl0617.html.

34. Federal Register Environmental Documents, "Interagency Cooperation under the Endangered Species Act," August 15, 2008; http://www.epa.gov/EPA-IMPACT/2008/August/Day-15/i18938.htm.

35. Hardin, "The Cybernetics of Competition," 78–90.

36. Arne Naess, *Ecology, Community, and Lifestyle: Outline of an Ecosophy* (Cambridge: Cambridge University Press, 1989), 56–57.

37. Arne Naess, "The Shallow and the Deep, Long-Range Ecology Movements: A Summary," in *Deep Ecology for the 21st Century*, ed. G. Sessions, 225–239 (Boston: Shambhala, 1995), 151.

38. William James, *A Pluralistic Universe* (Rockville, MD: Arc Manor, 2008), 129.

39. James Kay, "Framing the Situation: Developing a System Description," in *The Ecosystem Approach: Complexity, Uncertainty, and Managing for Sustainability*, ed. David Waltner-Toews, James Kay, and Nina-Marie Lister, 16 (New York: Columbia University Press, 2008).

40. Charles Darwin, *The Origin of Species* (Middlesex: Penguin, 1968 [1859]), 114.

41. Paul R. Ehrlich and Peter Raven, "Butterflies and Plants: A Study in Co-Evolution," *Evolution* 18 (1965): 586–608.

42. S. L. Rose, J. Kamin, and R. C. Lewontin, *Not in Our Genes: Biology, Ideology and Human Nature* (New York: Penguin, 1984), 272; R. C. Lewontin, "Organism and Environment" in *Learning, Development, and Culture: Essays in Evolutionary Epistemology*, ed. H. C. Plotkin, 160 (Hoboken, NJ: Wiley, 1982).

43. Fritjof Capra, *The Turning Point* (New York: Simon and Schuster, 1982), 16. See also Fritjof Capra, *The Web of Life: A New Scientific Understanding of Living Systems* (New York: Doubleday, 1996), 298, 301.

44. Michael E. Soulé, quoted in *Defenders* (published by Defenders of Wildlife), Fall 1994, 39.

45. Wendell Berry, *The Gift of Good Land* (San Francisco: North Point Press, 1981), ix, 116.

46. James Kay, "An Introduction to Systems Thinking," in *The Ecosytem Approach: Complexity, Uncertainty, and Managing for Sustainability*, ed. David Waltner-Toews, James Kay, and Nina-Marie Lister (New York: Columbia University Press, 2008), 8–9.

47. Jacques Leslie, "The Last Empire," *Mother Jones*, February 2008, 29–39.

48. Barry Commoner, *The Closing Circle* (New York: Knopf, 1971), 33.

49. Ibid., 39, 40.

50. Capra, *The Turning Point*, 16. See also Fritjof Capra, "Systems Theory and the New Paradigm," in *Ecology: Key Concepts in Critical Theory*, ed. Carolyn Merchant, 335 (Atlantic Highlands, NJ Humanities Press, 1994).

51. Quoted in Wes Jackson, "Toward an Ignorance-Based Worldview" in *The Virtues of Ignorance: Complexity, Sustainability, and the Limits of Knowledge*, ed. Bill Vitek and Wes Jackson, 26 (Lexington: University of Kentucky Press, 2008).

Chapter 2

1. Patrick Hughes, *Guess Who's Coming to Breakfast?* (Stoughton Mass.: Packard Manse Media, 1978).

2. See Sara J. Scherr and Sajal Sthapit, "Farming and Land Use to Cool the Planet," in Robert Engelman et al., *State of the World 2009* (New York: Norton, 2009), 30–49.

3. Larry Chang, *Wisdom for the Soul: Five Millennia of Prescriptions for Spiritual Healing* (Washington, DC: Gnosophia Publishers, 2006), 292.

4. Martin Luther King Jr., "Letter from Birmingham City Jail," in *A Testament of Hope: The Essential Writings and Speeches of Martin Luther King, Jr.*, ed. James Melvin Washington, 290 (San Francisco: Harper, 1991).

5. Aristotle, *Ethics*, trans. J.A.K. Thomson (New York: Penguin, 1953), 62.

6. See Joseph Dunne, *Back to the Rough Ground: "Phronesis" and "Techne" in Modern Philosophy and in Aristotle* (Notre Dame: University of Notre Dame Press, 2000), 242.

7. Aristotle, *Ethics*, 89.

8. Ibid., 65.

9. Ibid., 228, and see also 166.

10. Alexander Pope, *An Essay on Man*, vol. 2 of *The Works of Alexander Pope* (London: John Murray, 1871), 415.

11. Aristotle *Ethics*, 167.

12. Ibid., 57; see also 66.

13. Kant develops his ethical thought in three major works: *Groundwork of the Metaphysic of Morals* (1785), *Critique of Practical Reason* (1788), and *Metaphysics of Morals* (1797).

14. Aldo Leopold, *A Sand County Almanac, with Essays on Conservation from Round River* (New York: Ballantine Books, 1966), 239.

15. Ibid.

16. Ibid., 262.

17. Ibid., 263.

18. Immanuel Kant, *Grounding for the Metaphysics of Morals*, trans. James W. Ellington (New York: Hackett, 1993), 3.

19. Mahabharata 5:1517, quoted in Robert Kane, *Through the Moral Maze: Searching for Absolute Values in a Pluralistic World* (Armonk, NY: Sharpe, 1996), 34.

20. *Udânavarga: A Collection of Verses from the Buddhist Canon: Being the Northern Buddhist Version of Dhammapada*, ed. Prajnavarman Dharmatrata and William Woodville Rockhill, 27 (Delhi: Rare Reprints, 1982).

21. Confucius, *The Analects* (New York: Penguin, 1979), 135.

22. *The Babylonian Talmud*, ed. Isidore Epstein, 140 (London: Soncino Press, 1938). See also *Babylonian Talmud*, Tractate Shabbath, Folio 31a, http://www.come-and-hear.com/shabbath/shabbath_31.html.

23. Matthew 7:12, *The New Jerusalem Bible* (New York: Doubleday, 1985), 1620.

24. Francis H. Cook, *Hua-yen Buddhism: The Jewel Net of Indra* (University Park: Pennsylvania State University Press, 1977), 118–119.

25. William James, *The Will to Believe,* quoted in E. F. Schumacher, *A Guide for the Perplexed* (New York: HarperCollins, 1977), 60.

26. Aristotle, *Ethics*, 231.

27. Quoted in George Ainslie, *Breakdown of Will* (Cambridge: Cambridge University Press, 2001), 145.

28. Edward O. Wilson, *Biophilia* (Cambridge, MA: Harvard University Press, 1984), 139. See also Stephen Kellert and Edward O. Wilson, eds., *Biophilia Hypothesis* (Washington, DC: Island Press, 1993).

29. Leopold, *A Sand County Almanac*, 261.

30. Edmund Burke, *Reflections on the Revolution in France* (New York: Doubleday, 1961), 59.

31. Richard Alexander, *The Biology of Moral Systems* (New York: Aldine de Gruyter, 1987), 80.

32. Garrett Hardin, *The Limits of Altruism: An Ecologist's View of Survival* (Bloomington: Indiana University Press, 1977), 132–133.

33. Garrett Hardin, "Lifeboat Ethics: The Case against Helping the Poor," *Psychology Today*, September 1974, www.garretthardinsociety.org

34. Garrett Hardin, "The Survival of Nations and Civilization," *Science* 172 (1971):1297, quoted in Barry Commoner, *The Closing Circle* (New York: Knopf, 1971), 297.

35. Commoner, *The Closing Circle*, 297.

36. Hardin, *The Limits of Altruism*, 27, 42.

37. Martin Bunch, Dan McCarthy, and David Waltner-Toews, "A Family of Origin for an Ecosystem Approach to Managing for Sustainability, in *The Ecosystem Approach:*

Complexity, Uncertainty, and Managing for Sustainability, ed. David Waltner-Toews, James Kay, and Nina-Marie Lister, 123–138 (New York: Columbia University Press, 2008), 129.

38. Garrett Hardin, *Living within Limits* (New York: Oxford University Press, 1993), 199.

39. Niccolò Machiavelli, *The Prince* and *The Discourses* (New York: Modern Library), 51.

40. Ralph Waldo Emerson, from "Nature," in *Selected Writings of Emerson*, ed. Donald McQuade (New York: Modern Library, 1981), 23.

41. Frances Moore Lappé, *Getting a Grip 2* (Cambridge, MA: Small Planet Media, 2010), 112–113, 118.

Chapter 3

1. Hesiod, *Theogony* (New York: Penguin, 1973), 43.

2. Wendell Berry, "The Way of Ignorance," in *The Virtues of Ignorance: Complexity, Sustainability, and the Limits of Knowledge,* ed. Bill Vitek and Wes Jackson, 37 (Lexington: University of Kentucky Press, 2008).

3. Fred Pearce, "Bangladesh's Arsenic Poisoning: Who Is to Blame?" *UNESCO Courier*, http://www.unesco.org/ulis/cgi-bin/ulis.pl?catno=121516&set=4B8E6845_3_155&g p=0&lin=1&ll=1; Allen Smith, Elena Lingas, and Mahfuzar Rahman, "Contamination of Drinking-Water by Arsenic in Bangladesh: A Public Health Emergency," *Bulletin of the World Health Organization* 78, no. 9 (2000), 1093–1103.

4. Daniel P. Y. Chang, "Overview: Environmental Impacts of a Motor Fuel Additive: Methyl Tertiary Butyl Ether (MTBE)," in *Managing for Healthy Ecosystems*, ed. David Rapport, Bill L. Lasley, Dennis E. Rolston, N. Ole Nielsen, Calvin O. Qualset, Ardeshir B. Damania, 14–33 (Boca Raton: Lewis Publishers, 2003).

5. HFCs, the most commonly used of the F-gases that have replaced CFCs since the Montreal Protocol came into effect in 1989, have almost four thousand times the warming effect per pound of carbon dioxide. See Janos Mate, Kert Davies, and David Kanter, "The Risks of Other Greenhouse Gases," in *State of the World 2009*, ed. Robert Engelman et al. (New York: Norton, 2009), 52–55.

6. See Borlaugh's lecture, http://nobelprize.org/nobel_prizes/peace/laureates/1970/ borlaug-article.html, 21.

7. Union of Concerned Scientists, "Impacts of Genetically Engineered Crops on Pesticide Use in the United States: The First Thirteen Years," *Catalyst*, Spring 2010, 5.

8. Currently a team of researchers from the United States and Egypt has begun a three-year study to evaluate the balance sheet on Aswan's intended and unintended consequences. See "Aswan's Impact," *Time*, July 28, 2008, http://www.time.com/time/magazine/article/0,9171,913043,00.html. See also S. H. Sharaf el Din, "Effect of the Aswan High Dam on the Nile Flood and on the Estuarine and Coastal Circulation Pattern along the Mediterranean Egyptian Coast," *Limnology and Oceanography* 22 (1977): 194–207. Melvin A. Benarde, *Our Precarious Habitat* (Hoboken, NJ: Wiley, 1989), 18–20.

9. Paul Crutzen, A. R. Mosier, K. A. Smith, and W. Winiwarter, "N_2O release from Agro-Biofuel Production Negates Global Warming Reduction by Replacing Fossil Fuels," *Atmospheric Chemistry and Physics* 8 (2008): 389–395, http://www.atmos-chem-phys.net/8/389/2008/acp-8-389-2008.html.

10. Timothy Searchinger, Ralph Heimlich, R. A. Houghton, Fengxia Dong, Amani Elobeid, Jacinto Fabiosa, Simla Tokgoz, Dermot Hayes, and Tun-Hsiang Yu, "Use of U.S. Croplands for Biofuels Increases Greenhouse Gases through Emissions from Land Use Change," *Science Magazine*, February 7, 2008, http://www.sciencemag.org/cgi/content/abstract/1151861. Joseph Fargione, Jason Hill, David Tilman, Stephen Polasky, and Peter Hawthorne, "Land Clearing and the Biofuel Carbon Debt," *Science Magazine*, February 7, 2008. Ben Webster, "Green Fuels Cause More Harm Than Fossil Fuels, According to Report," *Times Online*, March 1, 2010, http://www.timesonline.co.uk/tol/news/environment/article7044708.ece.

11. "Study: Corn-Based Biofuel Costs 50 Gallons of Water per Mile," *Environmental Protection*, May 5, 2009, http://eponline.com/articles/2009/05/05/study-cornbased-biofuel-costs-50-gallons-of-water-per-mile.aspx.

12. C. Ford Runge and Benjamin Senauer, "How Biofuels Could Starve the Poor," *ForeignAffairs*, May/June 2007, http://www.foreignaffairs.com/articles/62609/c-ford-runge-and-benjamin-senauer/how-biofuels-could-starve-the-poor. D. Pimentel, "Biofuel Food Disasters and Cellulosic Ethanol Problems," *Bulletin of Science and Technology Society* 29 (2009): 205–214.

13. See "Excerpts from the Will of Alfred Nobel," http://nobelprize.org/alfred_nobel/will/short_testamente.html.

14. Martin Heidegger, *The Question Concerning Technology and Other Essays*, trans. W. Lovitt (New York: HarperCollins, 1977), 17.

15. Marshall McLuhan, "The Playboy Interview: Marshall McLuhan," *Playboy*, March 1969, http://www.playboy.com/articles/marshall-mcluhan-playboy-interview/index.html?page=2.

16. Bill McKibben, *The End of Nature* (New York: Random House, 1989), 210–217.

17. Quoted in Herman Daly, *Beyond Growth: The Economics of Sustainable Development* (Boston: Beacon Press, 1996), 17.

18. Wendell Berry, "Solving for Pattern," in *The Gift of Good Land* (San Francisco: North Point Press, 1981), 137–138.

19. Berry, *The Gift of Good Land*, 141, 144.

20. Gary Stix, "Turbocharging the Brain," *Scientific American* 301, no. 4 (2009): 46–55

21. The history of Polyface can be accessed at http://www.polyfacefarms.com/story .aspx.

22. Michael Pollan, *The Omnivore's Dilemma* (New York: Penguin, 2007), 209. See also "Michael Pollan Gives a Plant's-Eye View," a TED presentation at http://www .ted.com/index.php/talks/michael_pollan_gives_a_plant_s_eye_view.html.

23. Ibid., 213.

24. Ibid., 212.

25. Berry, *The Gift of Good Land*, 143.

26. E. F. Schumacher, *Small Is Beautiful: Economics as if People Mattered* (New York: HarperCollins, 1973), 35.

27. Ralf Yorque, Brian Walker, C. S. Holling, Lance Gunderson, Carl Folke, Stephen Carpenter, and William Brock, "Toward an Integrative Synthesis," in *Panarchy: Understanding Transformations in Human and Natural Systems,* ed. Lance H. Gunderson and C. S. Holling, 438 (Washington, DC: Island Press, 2002).

28. Michael Shermer, *The Mind of the Market* (New York: Holt, 2008), 259.

29. Jimmy Wales, "An Appeal from Wikipedia Founder, Jimmy Wales," http:// wikimediafoundation.org/wiki/Appeal/en?utm_source=2009_Jimmy_Appeal9& utm_medium=sitenotice&utm_campaign=fundraiser2009&referrer=http%3A%2F% 2Fen.wikipedia.org%2Fwiki%2FPlotinus&target=Appeal.

30. Jim Giles, "Internet Encyclopedias Go Head to Head," *Nature*, December 15, 2005, 900ff.

31. See Nicholas Carr, "Is Google Making Us Stupid?" *Atlantic* (July/August 2008), 56–63.

32. See E. Rogers, *Diffusion of Innovation* (New York: Free Press, 1995), 112–114.

33. James Hansen, "Paul Crutzen," *Time*, October 17, 2007, http://www.time.com/ time/specials/2007/article/0,28804,1663317_1663323_1669906,00.html.

34. Ken Caldeira, "Geoengineering to Shade Earth," in *State of the World 2009*, ed. Robert Engelman, Michael Renner, and Janet Sawin, 96–98 (New York: Norton, 2009). T. M. L. Wigley, "A Combined Mitigation/Geoengineering Approach to Climate Stabilization," *Science* 314 (2006): 452–454.

35. Paul J. Crutzen, "Albedo Enhancement by Stratospheric Sulfur Injections: A Contribution to Resolve a Policy Dilemma?" *Climatic Change* 77 (2006), 211–219. Wigley, "A Combined Mitigation/Geoengineering Approach to Climate Stabilization," 452. Jay Michaelson, "Geoengineering: A Climate Change Manhattan Project," *Stanford Environmental Law Journal* (January 1998), http://www.metatronics.net/lit/geo2.html#FN236.

36. Crutzen, "Albedo Enhancement by Stratospheric Sulfur Injections," 217.

37. "Top Scientists: Dirty Skies Could Help Keep Earth Cool," *Gainesville Sun*, November 17, 2006, 5A.

38. Graeme Wood, "Moving Heaven and Earth," *Atlantic* (July 2009), 73.

39. Thomas Lovejoy, "Climate Change's Pressures on Biodiversity," in *State of the World 2009*, ed. Robert Engelman, Michael Renner, and Janet Sawin, 67–70 (New York: Norton, 2009).

40. See John Vidal. "World Braced for Terminator 2," *Guardian*, October 6, 1999, http://www.guardian.co.uk/science/1999/oct/06/gm.food2.

41. The court ruling can be viewed at "Judgments of the Supreme Court of Canada," http://csc.lexum.umontreal.ca/en/2004/2004scc34/2004scc34.html.

42. Dough Gurian-Sherman and Emily Robinson, "Failure to Yield," *Catalyst*, Summer 2009, 11.

43. Bill McKibben, *Enough: Staying Human in an Engineered Age* (New York: Holt, 2003), 33–34.

44. Heidegger, *The Question Concerning Technology*, 32.

45. Martin Heidegger, *Discourse on Thinking* (New York: HarperCollins, 1966), 54.

46. Some scientists question whether the Chicxulub asteroid that hit the Yucatan was the direct and single cause of the demise of the dinosaurs and the mass extinction at the end of the Cretaceous period. See "The Chicxulub Debate," http://geoweb.princeton.edu/people/faculty/keller/chicxpage1.html

47. See "The Goal of the B612* Foundation Is . . . ," http://www.b612foundation.org/.

48. Arthur Koestler, *The Ghost in the Machine* (New York: Macmillan, 1967), 312.

Chapter 4

1. Dr. Seuss, *The Lorax* (New York: Random House, 1971).

2. For example, see http://www.seussville.com/loraxproject/.

3. Lester R. Brown, *Plan B 3.0: Mobilizing to Save Civilization* (New York: Norton, 2008), 283–284.

4. Jim MacNeill, Pieter Winsemius, and Taizo Yakushiji, *Beyond Interdependence: The Meshing of the World's Economy and the Earth's Ecology* (New York: Oxford University Press, 1991), 4.

5. Adam Smith, *The Wealth of Nations* (New York: Penguin, 1982), 119.

6. Rob Norton "Unintended Consequences," 1, http://www.econlib.org/LIBRARY/Enc/UnintendedConsequences.html.

7. Geiovann Di Chiro, "Indigenous Peoples and Biocolonianism," and J. Timmons Roberts, "Globalizing Environmental Justice," in *Environmental Justice and Environmentalism: The Social Justice Challenge to the Environmental Movement*, ed. Ronald Sandler and Phaedra Pezzulo, 251–283, 285–307 (Cambridge, MA: MIT Press, 2007), See also the Council for Responsible Genetics at http://www.councilforresponsiblegenetics.org/ and Action Bioscience at http://www.actionbioscience.org/.

8. Garrett Hardin, "The Tragedy of the Commons," *Science*, December 13, 1968, 1243–1248.

9. Aristotle, *The Politics* and *The Constitution of Athens* (Cambridge: Cambridge University Press, 1996), 33.

10. See Elinor Ostrom, "A General Framework for Analyzing Sustainability of Social-Ecological Systems," *Science*, July 24, 2009, 419–422. See also Elinor Ostrom, *Governing the Commons: The Evolution of Institutions for Collective Action* (Cambridge: Cambridge University Press, 1990).

11. Kenneth Boulding, "The Economics of the Coming Spaceship Earth," in *Environmental Quality in a Growing Economy*, H. Jarrett, ed., 3–14 (Baltimore, MD: Resources for the Future/Johns Hopkins University Press, 1966).

12. Buckminster Fuller, *Operator's Manual for Spaceship Earth* (Carbondale: Southern Illinois University Press, 1969).

13. Adlai Stevenson, "Strengthening the International Development Institutions," July 9, 1965, http://www.adlaitoday.org/ideas/connect_sub2_engage.html

14. Stevenson, "Strengthening the International Development Institutions."

15. Anup Shah, "Poverty Facts and Stats," *Global Issues*, updated March 2009, http://www.globalissues.org/article/26/poverty-facts-and-stats.

16. Quoted in Stephen Schmidheiny, *Changing Course: A Global Business Perspective on Development and the Environment* (Cambridge, MA: MIT Press, 1992), 135.

17. Vandana Shiva, *Earth Democracy: Justice, Sustainability, and Peace* (Boston: South End Press, 2005), 61.

18. See Brown, *Plan B 3.0*, 121–123.

19. David Brooks, "The Return of History," *New York Times*, March 26, 2010, A23.

20. Thomas Friedman, "The Great Iceland Meltdown," *New York Times*, October 18, 2008, http://www.nytimes.com/2008/10/19/opinion/19friedman.html.

21. Thomas L. Friedman, *The World Is Flat* (New York: Farrar, Straus and Giroux, 2006), 522.

22. Karl Marx and Friedrich Engels, "The Communist Manifesto," in *Marx/Engels Selected Works*, vol. 1 (Moscow: Progress Publishers, 1969), 98–137, quoted in ibid., 236.

23. See the *KOF Index of Globalization*, http://globalization.kof.ethz.ch/, and the aggregate data from 1970 available at http://globalization.kof.ethz.ch/query/. See also the A. T. Kearney/Foreign Policy Globalization Index at http://www.atkearney.com/index.php/Publications/globalization-index.html.

24. Michael Shermer, *The Mind of the Market* (New York: Holt, 2008), 22.

25. Friedman, *The World Is Flat*, 234.

26. Steven M. Beaudoin, *Poverty in World History* (London: Routledge, 2007), 98.

27. See the Gini Index of distribution of family income in the CIA's *World Factbook*, at https://www.cia.gov/library/publications/the-world-factbook/fields/2172.html. See also the 2006 edition of the A. T. Kearney/Foreign Policy Globalization Index.

28. See the Kearney/Foreign Policy Globalization Index.

29. Quoted in Colin Payne, "Recent National Trade Agreement Is a Bad Deal, says Evans," *Nelson Daily News*, July 2008, 3, 5.

30. Jacques Leslie, "The Last Empire," *Mother Jones* (February 2008): 29–39. Matt Jenkins, "A Really Inconvenient Truth," *Miller-McCune* (April–May 2008): 39–49.

31. George Bernard Shaw, *Man and Superman* (New York: Penguin, 1957), 221.

32. In 2001, Goldman Sachs coined the acronym BRIC economies (Brazil, Russia, India, and China) and suggested that these countries, accounting for 40 percent of the world's population and a quarter of its land, would surpass the combined

economies of the world's wealthiest nations by 2050. See http://www2
.goldmansachs.com/ideas/index.html.

33. Happy Planet Index 2.0, 8, http://www.happyplanetindex.org/public-data/files/
happy-planet-index-2-0.pdf.

34. E. F. Schumacher, *Small Is Beautiful: Economics as if People Mattered* (New York:
HarperCollins, 1973), 15, 19.

35. Ibid., 20.

36. Herman Daly, *Beyond Growth: The Economics of Sustainable Development* (Boston:
Beacon Press, 1996), 88.

37. Ibid., 78.

38. Ibid., 166.

39. Robert Costanza, Ralph d'Arge, Rudolf de Groot, Stephen Farber, Monica
Grasso, Bruce Hannon, Karin Limburg, Shahid Naeem, Robert V. O'Neill,
Jose Paruelo, Robert G. Raskin, Paul Sutton, and Marjan van den Belt, "The Value
of the World's Ecosystem Services and Natural Capital," *Nature* 387 (1997):
253–260.

40. John Seager, president of Population Connection (formerly Zero Population
Growth), observes that "news reports claim that the [Ponzi scheme] allegedly per-
petrated by Bernard L. Madoff is the biggest ever. Not even close. Population growth
remains king of the hill when it comes to the fraud named after one Charles Ponzi."
John Seager, *Reporter* 41, no. 1 (February 2009): 1.

41. Schumacher, *Small Is Beautiful*, 39.

42. As reported in Juliette Jowit, "World's Top Firms Cause $2.2tn of Environmental
Damage, Report Estimates," *Guardian*, February 18, 2010, http://www.guardian
.co.uk/environment/2010/feb/18/worlds-top-firms-environmental-damage.

43. See http://www.computertakeback.com/legislation/state_legislation.htm. For
the Canadian program, see http://www.ec.gc.ca/epr/default.asp?lang=En&n
=EEBCC813-1.

44. Paul Faeth, Robert Repetto, Kim Kroll, Qi Dai, and Glenn Helmers, *Paying the
Farm Bill* (Washington, DC: World Resources Institute, 1991).

45. Mike Nickerson, *Life, Money and Illusion: Living on Earth as if We Want to Stay*
(Lanark, Ontario: Seven Generations Publishing, 2006), 264. Frances Moore Lappé,
Getting a Grip 2 (Cambridge, MA: Small Planet Media, 2010), 80, 84.

46. See "Big Oil, Bigger Giveaways," Friends of the Earth, July 2008, http://www
.foe.org/pdf/FoE_Oil_Giveaway_Analysis_2008.pdf.

47. Robert Engelman, "Sealing the Deal to Save the Climate," in *State of the World 2009*, ed. Robert Engelman, Michael Renner, and Janet Sawin, 169–188 (New York: Norton, 2009).

48. Adam Smith, *An Inquiry into the Nature and Causes of the Wealth of Nations*, 2 vols. (Oxford: Clarendon Press, 1979), 625.

49. Quoted in Nickerson, *Life, Money and Illusion*, 158.

50. Quoted in Kai Ryssdal, "It's Too Easy Being Green," *Atlantic* (July/August 2010): 40.

51. Quoted in Nickerson, *Life, Money and Illusion*, 122.

52. See, for example, http://www.scorecard.org/.

53. See, for example, http://www.greenseal.org/ and www.goodguide.com.

54. See *Handbook of National Accounting: Integrated Environmental and Economic Accounting* (New York: United Nations, 2003), unstats.un.org/unsd/envAccounting/seea2003.pdf. See also the Human Development Reports compiled by the United Nations Development Program, http://hdr.undp.org/en/statistics/; the Calvert-Henderson Quality of Life Indicators, http://www.calvert-henderson.com/; and the Genuine Progress Indicator produced by Redefining Progress, http://www.solar783.com/solar783/gpi1999.pdf.

55. See the Happy Planet Index.

56. See "The Sustainability Report," http://www.sustreport.org/indicators/nrtee_esdi.html. See also the Canadian Index of Wellbeing developed by the Institute of Wellbeing, http://www.ciw.ca/en/TheCanadianIndexOfWellbeing.aspx.

57. See Pedro Conceição and Romina Bandura, "Measuring Subjective Wellbeing: A Summary Review of the Literature" (New York: Office of Development Studies, United Nations Development Programme Research Papers, May 2008), accessed at http://www.undp.org/developmentstudies/researchpapers.shtml.

58. Quoted in James Robertson, "Shaping the Post-Modern Economy," in *Business and the Environment*, ed. Richard Welford and Richard Starkey, 22 (Washington, DC: Taylor and Francis, 1996).

59. Francesca Lyman, "The Uholy Trinity," *Amicus Journal* (March 22, 1994): 8.

60. William Ophuls and Stephen Boyan Jr., *Ecology and the Politics of Scarcity Revisited: The Unraveling of the American Dream* (New York: Freeman, 1992), 226.

61. Carl. N. McDaniel, *Wisdom for a Livable Planet* (San Antonio, TX: Trinity University Press, 2005). 148.

62. Donald Murphy, "Our Need for Nature," *Amicus Journal* (Summer 1998): 30–31.

63. K. S. Shrader-Frechette, *Risk and Rationality: Philosophical Foundations for Populist Reforms* (Berkeley: University of California Press, 1991), 54.

64. Thomas Friedman, "The Price Is Not Right," *New York Times*, March 31, 2009, http://www.nytimes.com/2009/04/01/opinion/01friedman.html?_r=1.

65. Lappé, *Getting a Grip 2*, 183.

66. See Al Gore, *Earth in the Balance* (Boston: Houghton Mifflin, 1992), 170.

67. Ophuls and Boyan, *Ecology and the Politics of Scarcity Revisited*, 219. MacNeill, Winsemius, and Yakushiji, *Beyond Interdependence*, 46.

68. John S. Dryzek, *Rational Ecology: Environment and Political Economy* (New York: Basil Blackwell, 1987), 56.

69. David Sanger, "Long-Term Debt," *New York Times*, February 2, 2010, A1.

70. Edmund Burke, *Reflections on the Revolution in France* (New York: Doubleday, 1961), 108.

71. Daly, *Beyond Growth*, 31.

72. Paul Hawken, Amory Lovins, and L. Hunter Lovins, *Natural Capitalism* (New York: Little, Brown, 1999), 8.

73. Hawken, Lovins, and Lovins, *Natural Capitalism*, 14.

74. Daly, *Beyond Growth*, 15.

75. See, for example, J. Hartwick, "Intergenerational Equity and the Investing of Rents from Exhaustible Resources," *American Economic Review* 66 (1977): 972–74. See also Nickerson, *Life, Money and Illusion*.

76. Ralf Yorque, Brian Walker, C. S. Holling, Lance Gunderson, Carl Folke, Stephen Carpenter, and William Brock, "Toward an Integrative Synthesis," in *Panarchy: Understanding Transformations in Human and Natural Systems*, ed. Lance H. Gunderson and C. S. Holling, 438 (Washington, DC: Island Press, 2002).

77. John Dewey, *The Political Writings* (Indianapolis: Hackett, 1993), 171.

78. Shermer, *The Mind of the Market*, 140–46. See also R. Layard, G. Mayraz, and S. Nickell, "The Marginal Utility of Income," *Journal of Public Economics* 92 (2008): 1846–1857; Angus Deaton, "Income, Health and Wellbeing around the World: Evidence from the Gallup World Poll," *Journal of Economic Perspectives* 22 (2008): 53–77, http://www.ncbi.nlm.nih.gov/pmc/articles/PMC2680297/.

79. Shermer, *The Mind of the Market*, 147–148.

80. Ibid., 154. See also Kwame Anthony Appiah, *Experiments in Ethics* (Cambridge, MA: Harvard University Press, 2008).

81. Shermer, *The Mind of the Market*, 178.

82. Adam Smith, *A Theory of Moral Sentiments* (Amherst, MA: Prometheus Books, 2000), 3.

83. David Warsh, *On the Influence and Authority of Conscience (and Other Consider-ations Not Found in Any Economics Textbook*, in Economicprinciples.com, December 24, 2006, http://www.economicprincipals.com/issues/2006.12.24/229.html.

84. Smith, *A Theory of Moral Sentiments*, 56.

85. See Martin E. Seligman, *Authentic Happiness* (New York: Free Press, 2002).

86. Amory Lovins, interviewed in *Resurgence*, no. 198 (January/February 2000), http://www.resurgence.org/magazine/article1806-NATURAL-CAPITALISM.html. See also Hawken, Lovins and Lovins, *Natural Capitalism*, 261.

87. See John Holland, *Hidden Order: How Adaptation Builds Complexity* (Reading, MA: Addison-Wesley, 1995).

88. The Brundtland Report stipulates that "economic development is unsustainable if it increases vulnerability to crises." World Commission on Environment and Development, *Our Common Future* (New York: Oxford University Press, 1987), 53.

Chapter 5

1. Aristotle, *The Politics and The Constitution of Athens* (Cambridge: Cambridge University Press, 1996), 14.

2. Bernard Crick, *In Defense of Politics*, 4th ed. (Chicago: University of Chicago Press, 1992), 272.

3. See Rudolf Rummel, *Power Kills: Democracy as a Method of Nonviolence* (Rutgers, NJ: Transaction Publishers, 1997).

4. See Melvin Small and David J. Singer, "The War Proneness of Democratic Regimes, 1816–1965," *Jerusalem Journal of International Relations* 1 (1976): 50–69; Michael W. Doyle, "Kant, Liberal Legacies, and Foreign Affairs," *Philosophy and Public Affairs* 12 (1983): 205–235 , 323–353; James Lee Ray, *Democracy and International Conflict* (Columbia: University of South Carolina Press, 1995); James Lee Ray, "Does Democracy Cause Peace?" *Annual Review of Political Science* 1 (1998): 27–46.

5. See Jack Levy, "Domestic politics and war," *Journal of Interdisciplinary History* 18 (1988): 653–673.

6. See Steve Chan, "In Search of Democratic Peace: Problems and Promise," *Mershon International Studies Review* 41 (May 1997): 59–91.

7. Adolf G. Gundersen, *The Environmental Promise of Democratic Deliberation* (Madison: University of Wisconsin Press, 1995), 5, 159–165. See also John Barry,

"Sustainability, Political Judgement and Citizenship: Connecting Green Politics and Democracy," in *Democracy and Green Political Thought*, ed. Brian Doherty and Marius de Geus, 115–131 (London: Routledge, 1996).

8. See Hugh Ward, "Liberal Democracy and Sustainability," *Environmental Politics* 17 (June 2008): 386–409.

9. Reinhold Niebuhr, *Children of Light and Children of Darkness* (New York: Charles Scribner's Sons, 1944), xiii.

10. Harold Lasswell, *Politics: Who Gets What, When, and How* (New York: Meridian Books, 1958).

11. Thomas Hobbes, *Leviathan*, ed. C. B. Macpherson, 186 (New York. Penguin, 1968).

12. Plato, *Gorgias* (464b), in *The Collected Dialogues of Plato*, ed. Edith Hamilton and Huntington Cairns, 246 (Princeton, NJ: Princeton University Press, 1989).

13. Hannah Arendt, *The Human Condition* (Chicago: University of Chicago Press, 1958), 198.

14. I have grappled more extensively with this puzzle in Leslie Paul Thiele, "The Ontology of Action: Arendt and the Role of Narrative," *Theory and Event* 12, no. 4 (December 2009), http://muse.jhu.edu/journals/theory_and_event/.

15. Hannah Arendt, *Between Past and Future* (New York: Penguin Books, 1968), 152–53.

16. Ibid., 235.

17. Ibid., 164–165.

18. See Hannah Arendt, *The Life of the Mind* (New York: Harcourt, Brace, Jovanovich, 1978), 29, 32, 89, 200–207, 216.

19. Arendt, *The Human Condition*, 235.

20. Hannah Arendt, *The Origins of Totalitarianism* (Cleveland: World Publishing, 1958), 479; see also Arendt, *The Life of the Mind*, vol. 2:6, 217.

21. Hannah, Arendt, "On Hannah Arendt," in Melvyn A. Hill, *Hannah Arendt: The Recovery of the Public World* (New York: St. Martin's Press), 317.

22. See Patchen Markell, "The Rule of the People: Arendt, *Arche*, and Democracy," *American Political Science Review* 100 (February 2006): 1–14.

23. Arendt, "On Hannah Arendt," 317.

24. Arendt, *The Life of the Mind*, vol. 2: 33.

25. Arendt, *The Human Condition*, 192.

26. Hannah Arendt, *Men in Dark Times* (New York: Harcourt, Brace & World, 1968), 104.

27. Arendt, *The Human Condition*, 180.

28. Ibid., 184.

29. Arendt, "On Hannah Arendt," 303.

30. Arendt, *The Origins of Totalitarianism*, 479.

31. John Dewey, *Human Nature and Conduct* (Mineola, NY: Dover, 2002), 201.

32. Alasdair MacIntyre, *After Virtue* (Notre Dame: University of Notre Dame Press, 1981), 201.

33. Cicero, "On the Orator (I)," in *On the Good Life*, trans. Michael Grant, 316 (New York: Penguin, 1971).

34. Arendt, *Between Past and Future*, 219.

35. Ibid., 220.

36. Ralph Waldo Emerson, "Prudence," in *Selected Writings of Emerson*, ed. Donald McQuade, 222 (New York: Modern Library, 1981).

37. Mark Twain, *Mark Twain's Notebook*, ed. A. B. Paine, 344 (New York: Cooper Square Publishers, 1972).

38. Isaiah Berlin, *The Sense of Reality: Studies in Ideas and their History*, ed. Henry Hardy, 25 (London: Chatto and Windus, 1996). See also Ryan Patrick Hanley, "Political Science and Political Understanding: Isaiah Berlin on the Nature of Political Inquiry," *American Political Science Review* 98 (2004): 327–339.

39. Isaiah Berlin, *The Power of Ideas*, ed. Henry Hardy, 188 (Princeton, NJ: Princeton University Press, 2000).

40. Berlin, *The Sense of Reality*, 25.

41. La Rochefoucauld, *Maxims* (New York: Penguin, 1959), 67.

42. Quoted in G. V. Plekhanov, ed., *Fundamental Problems of Marxism* (London: Lawrence & Wishart, 1941), 55.

43. Aristotle, *Nichomachean Ethics* (1140b), trans. Terence Irwin, 154 (Indianapolis: Hackett Publishing, 1985).

44. Quoted in Jonathan Schell, *The Fate of the Earth* (New York: Avon, 1982), 109.

45. Quoted in Jay M. Shafritz, *The Harper Collins Dictionary of American Government and Politics* (New York: HarperCollins, 1993), 368.

46. Ambrose Bierce, *The Devil's Dictionary (1906)*, quoted in ibid., 369.

47. Arendt, *Between Past and Future*, 216; see Arendt, *The Human Condition*, 229.

48. See Joseph Schumpeter, *Capitalism, Socialism and Democracy*, 2nd ed. (New York: HarperCollins, 1950); Anthony Downs, *An Economic Theory of Democracy* (New York: HarperCollins, 1957).

49. Albert O. Hirschman, *Shifting Involvements: Private Interest and Public Action* (Princeton, NJ: Princeton University Press, 1982), 85–88.

50. Arendt, *The Human Condition*, 186.

51. *Bhagavad Gita*, trans. Stephen Mitchell, 54, 62, 65, 67, 82–83, 184 (New York: Three Rivers Press, 1988).

52. Arendt, *The Human Condition*, 206–207.

53. Sheldon Wolin, "What Revolutionary Action Means Today," in *Dimensions of Radical Democracy: Pluralism, Citizenship, Community*, ed. Chantal Mouffe, 251–252 (London: Verso, 1992).

54. Joseph Dunne, *Back to the Rough Ground: "Phronesis" and "Techne" in Modern Philosophy and in Aristotle* (Notre Dame: University of Notre Dame, 1993), 263.

55. Arendt, *The Human Condition*, 184.

Chapter 6

1. An unconscious placebo effect can occur in nonhuman animals. Rats demonstrate the same physiological effects from drinking an inactive, flavored beverage as they do from receiving a drug injection if, in previous days, they received a drug injection along with the inactive, flavored beverage. See Maj-Britt Niemi, "Cure in the Mind," *Scientific American Mind* (February–March 2009): 42–49.

2. Richard Dawkins, *The Ancestor's Tale: A Pilgrimage to the Dawn of Life* (London: Phoenix, 2005), 490.

3. Walt Whitman, *Leaves of Grass* (New York: Holt, Rinehart and Winston, 1966), 23.

4. David and Ann Premack, *Original Intelligence* (New York: McGraw-Hill, 2003), 139–157; Angeline Lillard and Lori Skibbe, "Theory of Mind: Conscious Attribution and Spontaneous Trait Inference," in *The New Unconscious*, ed. Ran Hassin, James Uleman, and John Bargh, 277–305 (New York: Oxford University Press, 2005); Bertram Malle, "Folk Theory of Mind: Conceptual Foundations of Human Social Cognition," in *The New Unconscious*, p. 229; Michael Tomasello, Joseph Call, and Brian Hare, "Chimpanzees Understand Psychological States–The Question Is Which Ones and to What Extent," *Trends in Cognitive Sciences* 7 (2003): 153–156.

5. Giacomo Rizzolatti and Laila Craighero, "The Mirror-Neuron System," *Annual Review of Neuroscience* 27 (2004):169–192.

6. Marco Iacoboni, *Mirroring People* (New York: Farrar, Straus and Giroux, 2008), 201.

7. Vittorio Gallese and Alvin Goldman, "Mirror Neurons and the Simulation Theory of Mind-Reading," *Trends in Cognitive Sciences* 2 (1998): 493–501.

8. Iacoboni, *Mirroring People*, p. 265.

9. Vittorio Gallese, "The 'Shared Manifold' Hypothesis: From Mirror Neurons to Empathy," *Journal of Consciousness Studies* 8 (2001): 33–50. The capacity for imitation and the capacity for emotional connection to other people are highly correlated. See Iacoboni, *Mirroring People*, 114.

10. Iacoboni, *Mirroring People*, 114, 268.

11. Norbert Schwarz, "Feelings as Information: Moods Influence Judgments and Processing Strategies," in *Heuristics and Biases: the Psychology of Intuitive Judgment*, ed. Thomas Gilovich, Dale Griffin and Daniel Kahneman, 534–535 (Cambridge: Cambridge University Press, 2002).

12. Nicholas Humphrey, *Seeing Red: A Study in Consciousness* (Cambridge, MA: Harvard University Press, 2006), 90, 98, 101.

13. Antonio Damasio, *Descartes' Error: Emotion, Reason, and the Human Brain* (New York: Penguin, 1994), 174, 184.

14. Joseph LeDoux, *The Synaptic Self: How Our Brains Become Who We Are* (New York: Penguin Books, 2002), ix, 2.

15. George Ainslie, *Breakdown of Will* (Cambridge: Cambridge University Press, 2001), 41, 43.

16. Michael S. Gazzaniga, *The Mind's Past* (Berkeley: University of California Press, 1998), 174. See also A. R. Damasio and H. Damasio, "Making Images and Creating Subjectivity," in *The Mind-Brain Continuum: Sensory Processes*, ed. Rodolfo Llinas and Patricia Churchland, 25 (Cambridge, MA: MIT Press, 1996); Antonio R. Damasio, *The Feeling of What Happens: Body and Emotion in the Making of Consciousness* (New York: Harcourt, Brace, 1999), 189.

17. Hermann Hesse, *Steppenwolf*, trans. Basil Creighton, 67 (New York: Bantam Books, 1963).

18. Hesse, *Steppenwolf*, 218.

19. Freud's tripartite division of the soul was complicated in his later writings by the introduction of two overarching drives: libido, or the life drive, and the death drive. Freud employed the names of the Greek gods of love, *Eros,* and death, *Thanatos,* for these two drives. They represent, respectively, the human tendency to seek

ever-greater satisfaction of manifold pleasures and the tendency to retreat from the sensual world and seek respite from its friction, pain, and change.

20. Friedrich Nietzsche, *Gesammelte Werke, Musarionausgabe* (Munich: Musarion Verlag, 1920–1929), 1:414, 7:395. See also Friedrich Nietzsche, *The Will to Power*, trans. Walter Kaufmann and R. J. Hollingdale, 211, 270 (New York: Vintage, 1968).

21. Friedrich Nietzsche, *Nietzsche contra Wagner: Out of the Files of a Psychologist*, in *The Portable Nietzsche* [1888], trans. Walter Kaufmann (New York: Viking, 1968).

22. Friedrich Nietzsche, *The Gay Science*, trans. Walter Kaufmann, 261, 262 (New York: Vintage, 1974).

23. Nietzsche, *The Will to Power*, 267. And see Nietzsche, *The Gay Science*, 215, 254, 261, 262.

24. See Leslie Paul Thiele, *Friedrich Nietzsche and the Politics of the Soul* (Princeton, NJ: Princeton University Press, 1990).

25. Friedrich Nietzsche, *On the Genealogy of Morals: A Polemic*, trans. Walter Kaufmann and R. J. Hollingdale, 121 (New York: Vintage, 1967). See also Nietzsche, *The Will to Power*, 149.

26. Aristotle, *The Politics*, ed. Stephen Everson, 16 (Cambridge: Cambridge University Press, 1996).

27. David Hume, *A Treatise of Human Nature* (London: Penguin Books, 1969), 462–65.

28. John Dewey, *Human Nature and Conduct* (Mineola, NY: Dover, 2002), 194, 196.

29. Ibid., 195–196.

30. Ibid.

31. Ibid., 95.

32. Ibid., 200.

33. Adam Smith, *A Theory of Moral Sentiments* (Amherst: Prometheus Books, 2000), 162.

34. Rose McDermott, "The Feeling of Rationality: The Meaning of Neuroscientific Advances for Political Science," *Perspectives on Politics* 2 (2004): 699. See also Damasio, *The Feeling of What Happens*, 41–42.

35. David E. Bell, Howard Raiffa, and Amos Tversky, "Descriptive, Normative, and Prescriptive Interactions in Decision Making," in *Decision Making: Descriptive, Normative, and Prescriptive Interactions*, ed. David Bell, Howard Raiffa, and Amos Tversky, 9 (Cambridge: Cambridge University Press, 1988.

36. Jung employed the word *psyche* to indicate the conscious and unconscious parts of the mind, reserving the world *Seele*, the German word for soul, as a reference for the conscious personality. In turn, he employed the term *Selbst*, or self, for the "sum total" of our consciousness and unconscious, the unified core of the well-integrated psyche. When the Romans translated the Greek, they employed the word *anima* for *psyche*. Jung employs *anima* to designate the "maternal eros." Carl Jung, "Phenomenology of the Self," in *The Portable Jung*, ed. Joseph Campbell, 152 (New York: Penguin Books, 1976). As valuable as it is, I do not abide by Jung's terminological lexicon.

37. C. G. Jung, *On the Nature of the Psyche*, trans. R.F.C. Hull (Princeton, NJ: Princeton University Press, 1960), 85–86.

38. Ibid., 95.

39. Ibid., 98. See also Carl Jung, "Instinct and the Unconscious," in *The Portable Jung*, 52.

40. Carl Jung, "Relation of Analytical Psychology to Poetry," in *The Portable Jung*, 319.

41. Carl Jung, "The Structure of the Psyche," in *The Portable Jung*, 45.

42. Jung, "Instinct and the Unconscious," 52. See also Carl Jung, "General Description of the Types," in *The Portable Jung*, 233.

43. Carl Jung, "Relations between the Ego and the Unconscious," in *The Portable Jung*, 119. See also Edward Edinger, *Ego and Archetype: Individuation and the Religious Function of the Psyche* (Boston: Shambhala, 1992), xiii, 103.

44. See David Tacey, *Jung and the New Age* (Philadelphia: Brunner-Routledge, 2001), 66.

45. Quoted in Curtis D. Smith, *Jung's Quest for Wholeness* (Albany: State University of New York Press), 73. Jung, *On the Nature of the Psyche*, 136. See also James Hillman, *Healing Fiction* (Barrytown, NY: Station Hill, 1983), 62–63; Smith, *Jung's Quest for Wholeness*, 62.

46. Jung, "Phenomenology of the Self," 145.

47. Ibid., 146.

48. See Luke 6:41–42 and Matthew 7:3–5.

49. Dewey, *Human Nature and Conduct*, 165.

50. Ibid., 156–157.

51. See Edinger, *Ego and Archetype*, 115.

52. Ibid., 116.

53. Aristotle, *The Nichomachean Ethics* (New York: Penguin Books, 1953), 62.

54. Ibid., 217.

55. LeDoux, *The Synaptic Self*, 73. And see Gerald Edelman, *Neural Darwinism* (New York: Basic Books, 1987).

56. Jon Kabat-Zinn, *Coming to our Senses: Healing Ourselves and the World through Mindfulness* (New York: Hyperion, 2005), 71.

57. Siddhārtha Gautama (Buddha) is reputed to have said: "All Beings are owners of their Karma. Whatever volitional actions they do, good or evil, of those they shall become the heir." Quoted in Jeffrey M. Schwartz and Sharon Begley, *The Mind and the Brain: Neuroplasticity and the Power of Mental Force* (New York: HarperCollins, 2002), 375.

58. See Edward Conze, *Buddhist Scriptures* (New York: Penguin, 1959), 139.

59. Francis H. Cook, *Hua-yen Buddhism: The Jewel Net of Indra* (University Park: Pennsylvania State University Press, 1977), 18–19.

60. Carl Jung, "The Spiritual Problem of Modern Man," in *The Portable Jung*, 470.

61. Schwartz, *The Mind and the Brain*, 164, 373. Daniel J. Siegel, *The Mindful Brain* (New York: Norton, 2007).

62. Schwartz, *The Mind and the Brain*, 325, 367.

63. Dewey, *Human Nature and Conduct*, 67, 72, 97, 105.

64. Ibid., 40.

65. Ibid., 38.

66. Ibid., 211.

67. Elkhonon Goldberg, *The Executive Brain: Frontal Lobes and the Civilized Mind* (New York: Oxford University Press, 2001), 218.

68. See Siegel, *The Mindful Brain*.

69. Damasio, *Descartes' Error*; Joseph LeDoux, *The Emotional Brain* (New York: Simon and Schuster, 1996).

70. Siegel, *The Mindful Brain*, 169.

71. Ibid., 192.

72. Hesse, *Steppenwolf*, 12.

73. Friedrich Nietzsche, *Thus Spoke Zarathustra: A Book for Everyone and No One*, trans. R. J. Hollingdale (New York: Penguin, 1969), 66.

74. Friedrich Nietzsche, *Daybreak: Thoughts on the Prejudices of Morality*, trans. R. J. Hollingdale (Cambridge: Cambridge University Press, 1982), 207.

75. See Leviticus 19:18. In the New Testament, the edict appears at Mark 12:31 and 12:33; Matthew 19:19 and 22:39; Luke 10:27, and frequently in the letters of St. Paul.

76. Whitman, *Leaves of Grass*, 40.

77. David Richo, *How to Be an Adult: A Handbook on Psychological and Spiritual Integration* (New York: Paulist Press, 1991), 57–58.

78. See Theodore Roszak, Mary E. Gomes, and Allen D. Kanner, *Ecopsychology: Restoring the Earth, Healing the Mind* (San Francisco: Sierra Club Books, 1995).

79. Theodore Roszak, *The Voice of the Earth: An Exploration of Ecopsychology* (New York: Touchstone, 1992), 19.

80. Roszak, *The Voice of the Earth*, 79, 63; Theodore Roszak, "Where Psyche Meets Gaia," 1–20, in *Ecopsychology*, 12.

81. Rinda West, *Out of the Shadow* (Charlottesville: University of Virginia Press, 2007), 191.

82. Jung, *On the Nature of the Psyche*, 118.

83. Edinger, *Ego and Archetype*, 103.

84. West, *Out of the Shadow*, 195.

85. Jung, "The Structure of the Psyche," 38.

86. David Abram, "The Ecology of Magic" in *Ecopsychology*, 305, 315.

87. James Hollis, *The Archetypal Imagination* (College Station: Texas A&M University Press, 2000), 57.

88. Dewey, *Human Nature and Conduct*, 252–53.

89. Abraham Maslow, *Toward a Psychology of Being*, 2nd ed. (New York: Van Nostrand, 1968), 122, 140.

Chapter 7

1. The word *metaphysics* was coined by the early compilers of Aristotle's notebooks and referred simply to the notes that followed after his works on physics.

2. Wendell Berry, *The Unsettling of America: Culture and Agriculture* (San Francisco: Sierra Club, 1986), 47.

3. Daisetz Suzuki, *Zen and Japanese Culture* (Princeton, NJ: Princeton University Press, 1959), 28.

4. Martin Heidegger, *What Is Philosophy?* trans. W. Kluback and J. Wilde. (London: Vision Press, 1958), 53.

5. Heidegger is referring here to Leibniz's understanding. Martin Heidegger, *The Metaphysical Foundations of Logic*, trans. M. Heim (Bloomington: Indiana University Press, 1984), 68. Importantly, Heidegger believes that the pre-Socratic understanding of *Logos* began to deteriorate with the rise of the Sophists, men who employed philosophical argument to gain worldly advantage. It is further undermined with the development of academic philosophy by Plato. In Heidegger's account, Platonic philosophy denies difference and diversity in its effort to champion identity, the uniform order of the cosmos.

6. See Simon de Bruxelles, "Plastic Duck Armada Is Heading for Britain after 15-Year Global Voyage," *Times*, June 28, 2007, http://www.timesonline.co.uk/tol/news/uk/article1996553.ece; see also "Friendly Floatee," http://en.wikipedia.org/wiki/Friendly_Floatees.

7. See F. Cramer, *Chaos and Order: The Complex Structure of Living Systems*, trans. D. I. Loewus (Weinheim: VCH, 1993).

8. Pierre-Simon Laplace, *A Philosophical Essay on Probabilities* (New York: Dover, 1951), 4.

9. Werner Heisenberg, *Physics and Philosophy* (New York: HarperCollins, 1971), 58.

10. David Bohm, *Wholeness and the Implicate Order* (London: Routledge, 1980), 105. When Heisenberg developed the uncertainty principle, he and fellow physicist Niels Bohr interpreted its meaning more generally. They developed the so-called Copenhagen interpretation, suggesting that the wave function should be understood less as a physical entity than as the product of mathematical abstraction, a tool for calculating the probability of the momentum or position of particles. For those who accept the Copenhagen interpretation, the wave function is not an ontological reality but an epistemological artifact. It pertains not to the way the universe really is (at the deepest level) but how we humans gain knowledge of it. The Copenhagen interpretation is pragmatist, even positivist, in maintaining that that which cannot be measured is meaningless. See Victor J. Stenger, *The Unconscious Quantum: Metaphysics in Modern Physics and Cosmology* (Amherst, NY: Prometheus Books, 1995), 103.

11. See Henry Stapp, *Mindful Universe: Quantum Mechanics and the Participating Observer* (New York: Springer, 2007), 26–27.

12. Amir D. Aczel, *Entanglement* (New York: Penguin, 2001), 21.

13. Lloyd Morgan, *Emergent Evolution* (London: Williams and Norgate, 1923).

14. See Stuart A. Kauffman, *Reinventing the Sacred* (New York: Basic Books, 2008), 24.

15. Aristotle, *Metaphysics* (New York: Columbia University Press, 1952), 178.

16. P. W. Anderson, "More Is Different: Broken Symmetry and the Nature of the Hierarchical Structure of Science," *Science* 177 (August 1972): 393–396.

17. See Roger Penrose, *Shadows of the Mind* (New York: Oxford University Press, 1994).

18. Anderson, "More Is Different," 393.

19. See Donella Meadows, *Thinking in Systems*, ed. Diana Wright (White River Junction, VT: Chelsea Green Publishing, 2008).

20. See Tor Norretranders, *The User Illusion*, trans. Jonathan Sydenham (New York: Viking, 1998), 356.

21. See Ilya Prigogine, *The End of Certainty: Time, Chaos, and the New Laws of Nature* (New York: Free Press, 1997).

22. Quoted in Frank Wilczek, *The Lightness of Being: Mass, Ether, and the Unification of Forces* (New York: Basic Books, 2008), 77.

23. Albert Einstein, Boris Podolsky, and Nathan Rosen, "Can Quantum-Mechanical Description of Physical Reality Be Considered Complete?" *Physical Review* 47 (May 1935): 777–780. Reprinted in John Archibald Wheeler, "Law without Law," in *Quantum Theory and Measurement*, ed. John Archibald Wheeler and Wojciech Hubert Zurek, 138–141 (Princeton, NJ: Princeton University Press, 1983).

24. Aczel, *Entanglement*, 188.

25. Henry Stapp, *Mind, Matter and Quantum Mechanics*, 2nd ed. (Berlin: Springer, 2004), 8.

26. T. D. Ladd, F. Jelezko, R. Laflamme, Y. Nakamura, C. Monroe, and J. L. O'Brien, "Quantum Computers," *Nature*, March 4, 2010, 45–53.

27. Wilczek, *The Lightness of Being*, 74.

28. Ibid., 231, 236–237.

29. Bohm, *Wholeness and the Implicate Order*, 124–125.

30. Stenger, *The Unconscious Quantum*, 289.

31. Bohm, *Wholeness and the Implicate Order*, 203.

32. Ibid., 149.

33. Ibid., 189.

34. Fritjov Capra, *The Tao of Physics* (Boston: Shambhala, 1975), 139. Capra's work, like that of Bohm, remains controversial.

35. Richard Feynman, quoted in Richard Dawkins, *Unweaving the Rainbow: Science, Delusion and the Appetite for Wonder* (Boston: Houghton Mifflin, 1998), 50.

36. William James, *A Pluralistic Universe* (Rockville, MD: Arc Manor, 2008), 128–129.

37. James, *A Pluralistic Universe*, 130–131.

38. Wilczek, *The Lightness of Being*, 121–122.

39. Ibid., 195.

40. John Archibald Wheeler, "Law without Law," 194.

41. Margaret J. Wheatley, *Leadership and the New Science* (San Francisco: Berrett-Koehler, 1999), 36.

42. Stenger, *The Unconscious Quantum*, 31.

43. C. G. Jung, *On the Nature of the Psyche*, trans. R.F.C. Hull (Princeton, NJ: Princeton University Press, 1960), 5, 139.

44. Edward Edinger, *Ego and Archetype: Individuation and the Religious Function of the Psyche* (Boston: Shambhala, 1992), 178.

45. Quoted in Stenger, *The Unconscious Quantum*, 33.

46. Arthur Koestler, *The Ghost in the Machine* (New York: Macmillan, 1967), 48.

47. Ibid., 103.

48. Albert Einstein, in *The New Quotable Einstein*, ed. Alice Calaprice (Princeton, NJ: Princeton University Press, 2005), 206.

Conclusion

1. Sophocles, *Oedipus the King* and *Antigone* in *Sophocles I* (Chicago: University of Chicago Press, 1954), 76, 170–171, 204.

2. John Dewey, *Human Nature and Conduct* (Mineola, NY: Dover, 2002), 86.

3. Robert Hughes, "The Phantom of Utopia," *Time*, November 6, 2000, http://www.time.com/time/magazine/article/0,9171,998420-2,00.html.

4. Dashka Slater, "Low-Carbon Diet," *Sierra* (March/April 2008): 24.

5. See Lance Gunderson, Garry Peterson, and C. S. Holling, "Practicing Adaptive Management in Complex Social-Ecological Systems," in *Complexity Theory for a Sustainable Future*, edited by Jon Norberg and Graeme Cumming, 225 (New York: Columbia University Press, 2008).

6. E. F. Schumacher, *Small Is Beautiful: Economics as If People Mattered* (New York: HarperCollins, 1973), 37.

7. Aldo Leopold, *A Sand County Almanac, with Essays on Conservation from Round River* (New York: Ballantine Books, 1966), 210.

8. Quoted in Craig Holdrege, "Can We See with Fresh Eyes?" in *The Virtues of Ignorance: Complexity, Sustainability, and the Limits of Knowledge*, ed. Bill Vitek and Wes Jackson, 332 (Lexington: University of Kentucky Press, 2008).

9. C. S. Holling and Lance H. Gunderson, "Resilience and Adaptive Cycles," in *Panarchy: Understanding Transformations in Human and Natural Systems*, ed. Lance H. Gunderson and C. S. Holling, 31 (Washington, DC: Island Press, 2002).

10. Ibid., 32.

11. Michel de Montaigne, *The Complete Essays of Montaigne*, trans. Donald Frame, 92 (Stanford: Stanford University Press, 1965).

12. From Nabokov, *Look at the Harlequins* (New York: McGraw-Hill, 1974), quoted in Ilya Prigogine, *The End of Certainty: Time, Chaos, and the New Laws of Nature* (New York: Free Press, 1997), 154.

13. Wendell Berry, "The Way of Ignorance," in *The Virtues of Ignorance*, 47.

14. Bill Vitek and Wes Jackson, "Taking Ignorance Seriously," in *The Virtues of Ignorance*, 1.

15. Paul Hawken, Amory B. Lovins, and L. Hunter Lovins, *Natural Capitalism: Creating the Next Industrial Revolution*, 1999), 316.

16. Quoted in E. F. Schumacher, *A Guide for the Perplexed* (New York: HarperCollins, 1977), 39.

Selected Bibliography

Arendt, Hannah. *The Human Condition*. Chicago: University of Chicago Press, 1958.

Arendt, Hannah. *Between Past and Future*. New York: Penguin, 1968.

Aristotle. *Ethics*. New York: Penguin, 1953.

Aristotle. *The Politics*. Cambridge: Cambridge University Press, 1996.

Berman, Morris. *The Reenchantment of the World*. Ithaca, NY: Cornell University Press, 1981.

Berry, Wendell. *The Gift of Good Land*. San Francisco: North Point Press, 1981.

Berry, Wendell. *The Unsettling of America*. San Francisco: Sierra Club, 1986.

Bhagavad Gita. Trans. Stephen Mitchell. New York: Three Rivers Press, 1988.

Bohm, David. *Wholeness and the Implicate Order*. London: Routledge and Kegan Paul, 1980.

Brown, Lester. *Plan B 3.0*. New York: Norton, 2008.

Burke, Edmund. *Reflections on the Revolution in France*. New York: Doubleday, 1961.

Capra, Fritjof. *The Web of Life*. New York: Doubleday, 1996.

Carson, Rachel. *Silent Spring*. Boston: Houghton Mifflin, 1962.

Commoner, Barry. *The Closing Circle*. New York: Knopf, 1971.

Cook, Francis H. *Hua-yen Buddhism*. University Park: Pennsylvania State University Press, 1977.

Daly, Herman. *Beyond Growth*. Boston: Beacon Press, 1996.

Damasio, Antonio. *Descartes' Error*. New York: Penguin, 1994.

Dewey, John. *The Political Writings*. Indianapolis: Hackett, 1993.

Dewey, John. *Human Nature and Conduct*. Mineola, NY: Dover, 2002.

Dryzek, John S. *Rational Ecology*. New York: Basil Blackwell, 1987.

Dunne, Joseph. *Back to the Rough Ground*. Notre Dame: University of Notre Dame Press, 2000.

Edelman, Gerald. *Neural Darwinism*. New York: Basic Books, 1987.

Edinger, Edward. *Ego and Archetype*. Boston: Shambhala, 1992.

Friedman, Thomas L. *The World Is Flat*. New York: Farrar, Straus and Giroux, 2006.

Giddens, Anthony. *The Constitution of Society*. Berkeley: University of California Press, 1984.

Gunderson, Lance H., and C. S. Holling, eds. Panarchy. Washington, DC: Island Press, 2002.

Hardin, Garrett. *The Limits of Altruism*. Bloomington: Indiana University Press, 1977.

Hardin, Garrett. *Living within Limits*. New York: Oxford University Press, 1993.

Hawken, Paul, Amory Lovins, and L. Hunter Lovins. *Natural Capitalism*. New York: Little, Brown, 1999.

Heidegger, Martin. *What Is Called Thinking*. New York: HarperCollins, 1968.

Heidegger, Martin. *The Question Concerning Technology and Other Essays*. New York: HarperCollins, 1977.

Hesiod. *Theogony*. New York: Penguin Books, 1973.

Hesse, Hermann. *Steppenwolf*. New York: Bantam Books, 1963.

Hirschman, Albert O. *Shifting Involvements*. Princeton, NJ: Princeton University Press, 1982.

Hobbes, Thomas. *Leviathan*. New York: Penguin, 1968.

Hollis, James. *The Archetypal Imagination*. College Station: Texas A&M University Press, 2000.

Iacoboni, Marco. *Mirroring People*. New York: Farrar, Straus and Giroux, 2008.

James, William. *A Pluralistic Universe*. Rockville, MD: Arc Manor, 2008.

Jung, Carl. *On the Nature of the Psyche*. Princeton, NJ: Princeton University Press, 1960.

Jung, Carl. *The Portable Jung*. New York: Penguin Books, 1976.

Kabat-Zinn, Jon. *Coming to Our Senses*. New York: Hyperion, 2005.

Kant, Immanuel. *Grounding for the Metaphysics of Morals*. New York: Hackett, 1981.

Kauffman, Stuart A. *Reinventing the Sacred*. New York: Basic Books, 2008.

Koestler, Arthur. *The Ghost in the Machine*. New York: Macmillan, 1967.

Lasswell, Harold. *Politics: Who Gets What, When, and How*. New York: Meridian Books, 1958.

LeDoux, Joseph. *The Synaptic Self*. New York: Penguin Books, 2002.

Leopold, Aldo. *A Sand County Almanac*. New York: Ballantine Books, 1966.

Machiavelli, Niccolò. *The Prince and the Discourses*. New York: Modern Library, 1950.

MacNeill, Jim, Pieter Winsemius, and Taizo Yakushiji. *Beyond Interdependence*. New York: Oxford University Press, 1991.

Mauro, Filippo, Stephane Dees, and Warwick McKibben, eds. Globalisation, Regionalism, and Economic Interdependence. Cambridge: Cambridge University Press, 2008.

McKibben, Bill. *The End of Nature*. New York: Random House, 1989.

McKibben, Bill. *Enough*. New York: Holt, 2003.

Meadows, Donella. *Thinking in Systems*. Edited by Diana Wright. White River Junction, VT: Chelsea Green Publishing, 2008.

Naess, Arne. *Ecology, Community and Lifestyle*. Cambridge: Cambridge University Press, 1989.

Nickerson, Mike. *Life, Money and Illusion*. Lanark, Ontario: Seven Generations Publishing, 2006.

Nietzsche, Friedrich. *Beyond Good and Evil*. New York: Penguin, 1972.

Nietzsche, Friedrich. *Daybreak*. Cambridge: Cambridge University Press, 1982.

Norberg, Jon, and Graeme Cumming, eds. Complexity Theory for a Sustainable Future. New York: Columbia University Press, 2008.

Norretranders, Tor. *The User Illusion*. New York: Viking, 1998.

Ophuls, William, and Stephen Boyan, Jr. *Ecology and the Politics of Scarcity Revisited*. New York: Freeman, 1992.

Ostrom, Elinor. *Governing the Commons*. Cambridge: Cambridge University Press, 1990.

Penrose, Roger. *Shadows of the Mind*. New York: Oxford University Press, 1994.

Pimm, Stuart. *The Balance of Nature?* Chicago: University of Chicago Press, 1991.

Pinchot, Gifford. *The Fight for Conservation*. New York: Doubleday, 1910.

Pollen, Michael. *The Omnivore's Dilemma*. New York: Penguin, 2007.

Roszak, Theodore, Mary Gomes, and Allen Kanner, eds. Ecopsychology. San Francisco: Sierra Club Books, 1995.

Schmidheiny, Stephan. *Changing Course*. Cambridge, MA: MIT Press, 1992.

Schumacher, E. F. *Small Is Beautiful*. New York: HarperCollins, 1973.

Schwartz, Jeffrey M., and Sharon Begley. *The Mind and the Brain*. New York: Harper-Collins, 2002.

Sessions, George, ed. Deep Ecology for the 21st Century. Boston: Shambhala, 1995.

Seuss, Dr. *The Lorax*. New York: Random House, 1971.

Shermer, Michael. *The Mind of the Market*. New York: Holt, 2008.

Shiva, Vandana. *Earth Democracy*. Boston: South End Press, 2005.

Shrader-Frechette, K. S. *Risk and Rationality*. Berkeley: University of California Press, 1991.

Siegel, Daniel J. *The Mindful Brain*. New York: Norton, 2007.

Smith, Adam. *An Inquiry into the Nature and Causes of the Wealth of Nations*. Oxford: Clarendon Press, 1979.

Stapp, Henry. *Mindful Universe*. New York: Springer, 2007.

Stenger, Victor J. *The Unconscious Quantum*. Amherst, NY: Prometheus Books, 1995.

Thiele, Leslie Paul. *Friedrich Nietzsche and the Politics of the Soul*. Princeton, NJ: Princeton University Press, 1990.

Thiele, Leslie Paul. *Environmentalism for a New Millennium*. New York: Oxford University Press, 1999.

Thiele, Leslie Paul. *Timely Meditations: Martin Heidegger and Postmodern Politics*. Princeton, NJ: Princeton University Press, 1995.

Thiele, Leslie Paul. *The Heart of Judgment*. Cambridge: Cambridge University Press, 2006.

Van Driescshe, Jason, and Roy Van Driescshe. *Nature Out of Place*. Washington, DC: Island Press, 2000.

Vitek, Bill, and Wes Jackson, eds. The Virtues of Ignorance. Lexington: University of Kentucky Press, 2008.

Waltner-Toews, David, James Kay, and Nina-Marie Lister, eds. The Ecosystem Approach. New York: Columbia University Press, 2008.

Wapner, Paul. *Living Through the End of Nature*. Cambridge, MA: MIT Press, 2010.

Whitman, Walt. *Leaves of Grass*. New York: Holt, 1966.

Wilczek, Frank. *The Lightness of Being*. New York: Basic Books, 2008.

Wilson, Edward O. *Biophilia*. Cambridge, MA: Harvard University Press, 1984.

Worster, Donald. *Nature's Economy*. Cambridge: Cambridge University Press, 1994.

Index